Sensors and Transducers

Other Macmillan titles of related interest

New Electronics Series
Series Editor: Paul A. Lynn

Sensors and Transducers

Characteristics, Applications, Instrumentation, Interfacing

M.J. Usher and D.A. Keating

Department of Cybernetics
University of Reading

Second Edition

MACMILLAN

First edition 1985
Reprinted 1990
Second edition 1996

Published by
MACMILLAN PRESS LTD
Houndmills, Basingstoke, Hampshire RG21 6XS
and London
Companies and representatives
throughout the world

ISBN 0–333–60487–3

A catalogue record for this book is available
from the British Library

Printed in Great Britain by
Antony Rowe Ltd, Chippenham, Wilts

Contents

Preface

Most quantities that we need to measure are inherently analogue. There is nothing very digital about a length or a temperature and although light may be considered to consist of photons most measurements involve such large numbers that the process is effectively analogue. Our own senses are analogue so it is hardly surprising to find that the vast majority of physical sensors are also analogue. It is only since the developments in microprocessor technology that digital transducers have become important, and have sometimes captured an undue proportion of attention; however, they still have to measure analogue quantities and most digital transducers therefore employ exactly the same physical principles as their analogue counterparts. This book discusses most of the transducers in current use, whether digital or analogue. The coverage is primarily from the measurand standpoint; for example, the different types of length transducer are discussed and compared together in one chapter, although we have also included chapters summarising the various transducer technologies, such as solid-state or fibre-optic devices. Digital transducers are dealt with in the same way, under length or temperature as appropriate, but a chapter is also devoted to their summary and classification.

The words 'sensor' and 'transducer' are widely used in referring to sensing devices, the former having gained in popularity in recent years. This is a pity because 'transducer' stresses the change in form of energy basic to the sensing process and leads to an elegant and powerful classification of devices. The word 'transducer' is used here when considering a complete sensing device, in which there is bound to be a change in form of energy; the word 'sensor' is reserved for devices which 'respond to a stimulus' but are not energy converting, such as a thermistor, which simply changes its resistance in response to temperature.

The most important transducer parameters are 'responsivity' and 'detectivity'; the former refers to the response of a transducer to the applied measurand and the latter to the least input measurand that can be detected. The use of these two separate words removes the ambiguity of the word 'sensitivity', unfortunately sometimes used in describing transducers, which may refer to the response either to the desired input or to an undesired input, or both together. These two responses must be distinguished and this is correctly accomplished by 'responsivity' and 'detectivity', which are used throughout the book.

The aim of the book is to provide an integrated account of the principles and properties of the most important types of physical transducer. The first chapter discusses the types of physical energy and the corresponding signals, and identifies the three basic types of transducers: self-generators, modulators and modifiers. A synthesis of the subject is attempted in chapter 2, describing the analogies that exist between different types of physical system and showing how our understanding can sometimes be improved by considering an analogy of a particular device or circuit. Chapter 3 starts with a three-dimensional representation of all possible transducers and goes on to consider the basic physical mechanisms available for transduction. Chapter 4 develops the relevant expressions for amplifiers and transducer bridges that are required before the detailed descriptions of the basic transducers for length, temperature and light are given in chapters 5, 6 and 7. For each of these quantities the physical background and measurement standards are first explained, followed by both a theoretical treatment of the basic transducers and a description of their practical design and application. Chapter 8 includes the application of the basic transducers to several important fields of measurement, such as acceleration, force, pressure, flow and sound. Although transducers are usually thought of as input devices, output transducers are important in measurement systems, being usually referred to as 'actuators', and chapter 9 is devoted to the various types available. Chapter 10 is concerned with measurement systems, showing how transducers and actuators are used in complete systems, and including solid-state sensors, resonator sensors, optical fibre sensors, pyrometry and ultrasonics. The final chapter gives a summary and classification of digital transducers and an introduction to interfacing to computer systems. Many worked examples are given, together with a set of exercises at the end of each chapter, full solutions being provided at the end of the book.

The first edition of this book was devoted specifically to transducers and their characteristics, but in the second edition the authors have extended the coverage to include both instrumentation, in chapter 10, and digital transducers and interfacing, in chapter 11. The book is therefore subtitled 'Characteristics, Applications, Instrumentation, Interfacing'. It is intended as a basic undergraduate text for students in engineering, physics and information technology.

Acknowledgements

The authors wish to thank the former and present heads of the Cybernetics Department, Peter Fellgett and Kevin Warwick, for their encouragement and suggestions regarding the lecture courses on which the book is based, and Christine Usher who did much of the typing and corrections. They also acknowledge the feedback from the many Cybernetics and Engineering students who acted as willing guinea-pigs* during the development of the courses.

---

*Guinea-pigs are nocturnal animals that mostly sleep during the day.

1 Introduction

1.1 Analogue and digital quantities

Recent developments in technology and the availability of cheap microprocessors have led to an increased interest in sensing devices, particularly so-called digital devices suitable for direct interfacing to computer systems. Unfortunately (or perhaps fortunately), the only thing at all digital about human beings is that most of us have ten fingers. We are analogue animals living in an analogue world. The quantities we need to measure are inherently analogue; they can in principle take a continuous range of values, though we may prefer to round the values to whole numbers at some stage. There is nothing very digital about a length or a temperature, and although matter is discrete it is certainly not so to our senses and not so to the vast majority of our sensors. In fact it is quite difficult to think of anything in nature that is inherently digital; almost the only example in measurement is in counting numbers of particles (photons, γ-rays, etc.).

Although we may wish to use 'digital' sensors in our computer systems, it is important to realise that they are bound to employ exactly the same physical principles as their analogue counterparts. One sometimes reads of industry lamenting the lack of 'absolute digital sensors' and of the importance of effort directed towards their development. In fact, if one examines a typical 'digital sensor' one usually finds it to be a totally analogue device operated in such a way as to produce a digital output. For example, an optical encoder for angle measurement consists of a disc with opaque and transparent sections, as in figure 1.1(a), producing an output as in figure 1.1(b). The process is perfectly analogue but the output is simply digitised into two levels by a comparator circuit.

Similarly, a 'digital' sensing device producing a frequency or pulse train proportional to the quantity being sensed is misleadingly named; the measurement becomes digital only when the waveform produced is converted into digits by, for example, a counting circuit. In principle this is exactly similar to digitising an 'analogue' waveform (whose voltage is proportional to the quantity sensed) by an analogue-to-digital converter.

A search for 'absolute digital sensors' is therefore misdirected in principle. What is important is not whether we are using a particular physical property to produce a digital or analogue device, but what the property is and how we can best use it. We need to search instead for suitable physical principles – analogue ones!

1

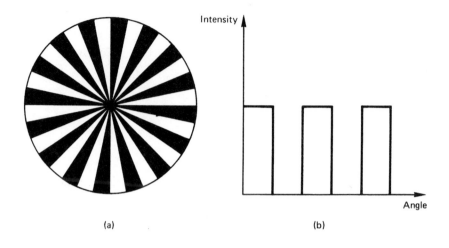

Figure 1.1 *Incremental optical encoder and output signal.*

This book will concentrate on understanding the basic physical principles available for use in sensing devices. We will see that by a sensible approach to the matter the classifications 'analogue' or 'digital' become hardly necessary, being simply a matter of method of operation, and that it is the underlying physical principles that are all-important.

1.2 Classification of sensing devices

In discussing sensing devices one has to decide whether to classify them according to the physical property they use (such as piezoelectric, photovoltaic, etc.) or according to the function they perform (such as measurement of length, temperature, etc.). In the former case one can present a reasonably integrated view of the sensing process, but it is a little disconcerting when one wishes to compare the merits of, say, two types of temperature sensor, if one has to look through separate sections on resistive, thermoelectric and semiconductor devices to make the comparison. Alternatively, books classifying devices by function often tend to be a rather boring catalogue of numerous unrelated devices. We will try to make a compromise here by first presenting an integrated view of the sensing process in terms of the way in which signals are transformed from one form to another. We will then further synthesise the subject by considering the analogies that exist between the different types of physical system, the various physical properties available for use in sensors and the relations between them.

Finally, we will be in a position to discuss sensing devices from the functional viewpoint, under headings such as length, temperature, etc., the prior discussion hopefully avoiding the appearance of an interminable list, yet suitable for someone who actually wants to select or use a sensor for a particular application rather than just read around the subject.

1.3 Sensors, transducers and actuators

The words 'sensor' and 'transducer' are both widely used in the description of measurement systems. The former is popular in the USA whereas the latter is more often used in Europe. The choice of words in science is rather important. In recent years there has been a tendency to coin new words or to misuse (or mis-spell) existing words, and this can lead to considerable ambiguity and misunderstanding, and tends to diminish the preciseness of the language. The matter has been very apparent in the computer and microprocessor areas, where preciseness is particularly important, and can seriously confuse persons entering the subject.

The word 'sensor' is derived from *sentire*, meaning 'to perceive' and 'transducer' is from *transducere* meaning 'to lead across'. A dictionary definition (*Chambers Twentieth Century*) of 'sensor' is 'a device that detects a change in a physical stimulus and turns it into a signal which can be measured or recorded'; a corresponding definition of 'transducer' is 'a device that transfers power from one system to another in the same or in different form'.

A sensible distinction is to use 'sensor' for the sensing element itself, and 'transducer' for the sensing element plus any associated circuitry. For example, a thermistor is a sensor, since it responds to a stimulus (changes its resistance with temperature), but only becomes a transducer when connected in a bridge circuit to convert change in resistance to change in voltage, since the complete circuit then transduces from the thermal to the electrical domain. A solar cell is both a sensor and a transducer, since it responds to a stimulus (produces a current or voltage in response to radiation) and also transduces from the radiant to the electrical domain. It does not require any associated circuitry, though in practice an amplifier would usually be used. All transducers thus contain a sensor, and many (though not all) sensors are also transducers.

We will use this convention here, though of course the distinction is rather small and as soon as one actually uses a sensor (by applying power to it) it becomes a transducer. An interesting classification of devices can be achieved by considering the various forms of energy or signal transfer and the word 'transducer' will be used most often in this book. Figure 1.2 shows the sensing process in terms of energy conversion.

The word 'actuate' means 'to put into, or incite to, action' and an actuator is a device that produces the display or observable output in a measurement system such as a light-emitting diode (LED) or moving coil meter. It is of course simply

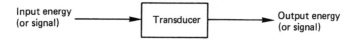

Figure 1.2 *The sensing process.*

a transducer used for output purposes, since it transduces from one domain to another (electrical to radiant for an LED).

1.4 Types of transducer

Since the conversion of energy from one form to another is an essential characteristic of the sensing process, it is useful to consider the various forms of energy. The following ten forms shown in table 1.1 have been distinguished (van Dijck, 1964).

Table 1.1   The main forms of energy

Type of energy	Occurrence
Radiant	Radio waves, visible light, infra-red, etc.
Gravitational	Gravitational attraction
Mechanical	Motion, displacement, forces, etc.
Thermal	Kinetic energy of atoms and molecules
Electrical	Electric fields, currents, etc.
Magnetic	Magnetic fields
Molecular	Binding energy in molecules
Atomic	Forces between nucleus and electrons
Nuclear	Binding energy between nuclei
Mass energy	Energy given by $E = mc^2$

It is often useful to think in terms of the types of signal associated with the various forms of energy. The essential characteristic of a signal is that of change as a function of time or space, since information cannot be carried or transmitted if a quantity remains constant. Each of the forms of energy has a corresponding signal associated with it, and for measurement purposes six types of signal are important:

(i)   Radiant: especially visible light or infra-red.
(ii) Mechanical: displacement, velocity, acceleration, force, pressure, flow, etc.
(iii) Thermal: temperature, heat flow, conduction, etc.
(iv) Electrical: voltage, current, resistance, dielectric constant, etc.
(v)   Magnetic: magnetic flux, field strength, etc.
(vi) Chemical: chemical composition, pH value, etc. (this does not appear in table 1, but is clearly a distinct form of signal, being derived from several forms of energy).

A general form of transducer system is shown in figure. 1.3 (after Middelhoek and Noorlag, 1981b). The signal is fed to an input transducer, which changes the form of energy, usually into electrical. The block labelled 'Modifier' represents an amplifier or other device that operates on the transduced signal, and an actuator (output transducer) then converts the energy into a form suitable for display or recording.

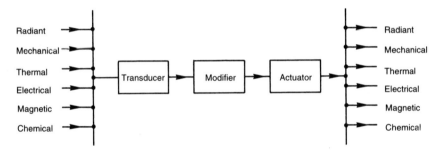

Figure 1.3   *A general transducer system.*

For example, the temperature of a hot body (thermal energy) could be measured by a thermocouple (input transducer) feeding an operational amplifier (modifier) followed by an LED display (actuator) producing radiant energy. Similarly, in a photographic exposure meter radiant energy falls on a photoconductive cell, producing a change in resistance; this change is converted to a voltage by a bridge circuit and amplifier, and produces a deflection of a meter. The overall transduction is between radiant and mechanical energy. A rather different action occurs in a bellows-type pressure gauge; the ambient pressure in the form of mechanical energy produces a deflection of the bellows which may be detected by a lever system, the transduction then being between two types of mechanical energy (pressure and displacement) rather than between two different forms of energy.

These three examples illustrate several important points. As mentioned above, we usually want our input transducer to produce an electrical signal that can conveniently be amplified or processed in some way, so the signals on either side of the modifier are usually electrical. Also, some transducers are said to be self-exciting or self-generating, in the sense that their operation does not require the application of external energy, whereas others, known as 'modulating transducers', do require such a source of energy. A thermocouple is self-generating, producing an e.m.f. in response to temperature difference, whereas a photoconductive cell is modulating. Without an external source of energy a photoconductive cell (effectively a light-dependent resistor) simply responds to the input light energy but does not produce a usable signal; a signal is obtained by applying an electrical voltage and monitoring the resulting current. Self-generators are sometimes referred to as 'passive' transducers and modulators as 'active' transducers. A third type of transducer is known as a modifier, and is characterised by the same form of energy at both input and output, as in the pressure gauge above. The same word is used in figure 1.3 because the energy form on each side of the modifier is electrical.

Self-generating transducers (thermocouples, piezoelectric, photovoltaic) usually produce very low output energy, having a low effective conversion efficiency; they are often followed by amplifiers to increase the energy level to a suitable value (to drive a meter, for example). In contrast, in modulating transducers, such as photoconductive cells, thermistors or resistive displacement devices, a relatively large flow of electrical energy is controlled by a much smaller input signal energy. Modifying transducers, such as elastic beams or diaphragms, may have a very high conversion efficiency between input and output energy, but usually require some other form of transducer to produce the required electrical output.

The three types of basic transducers are illustrated in figure 1.4.

1.5 Transducer parameters

The two most important parameters of a transducer are the output signal produced in response to a given input signal, and the output noise level. The word 'sensitivity' has often been used in connection with transducers, but unfortunately is rather ambiguous. For example, an instrument may be said to have high 'sensitivity' if it produces a large output in response to a given input, or because it is disturbed by people jumping up and down near it, or because it can detect very small input signals. These properties are quite distinct and require separate words to describe them.

The response of a transducer to an input signal is known as its responsivity r, defined by

$$r = \frac{\text{output signal in response to input}}{\text{input signal}}$$

For a displacement transducer the units would be V/m. The definition may be

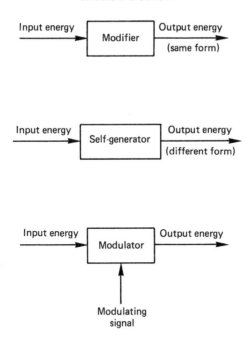

Figure 1.4 *The three types of transducer.*

applied between any chosen terminals of the device, so *r* may be in $\Omega/°C$ for a thermistor or $V/°C$ at the output of a bridge circuit connected to the thermistor.

In practice the responsivity may vary over the range of operation of the transducer, as illustrated in figure 1.5, which shows the output voltage as a function of displacement (curve OMZ) for a transducer of range 1 mm. (The figure is for a variable separation capacitive displacement transducer, described in detail in chapter 5.)

The responsivity is given by the slope of the curve and is constant and of value XY/OX for small displacements but rises as the displacement increases. With respect to the zero-displacement value, the maximum non-linearity at displacement 1 mm is YZ/XZ, expressed as a fraction of full scale. For non-linear transducers the output V_0 as a function of displacement *d* is often expressed in the general form

$$V_0 = a_0 + a_1 d + a_2 d^2 + a_3 d^3 + \ldots$$

where a_0 is an offset (very small in figure 1.5) and a_1 is the zero-displacement responsivity.

This type of transducer is often used as a null detector in force feedback systems and the appropriate responsivity would then be XY/OX with very low non-linearity. However, if used for displacement measurement over its full range

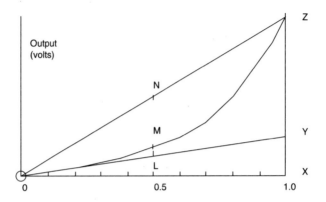

Figure 1.5 *Response of non-linear transducer.*

of 1 mm, the responsivity would usually be quoted as the 'mean' responsivity XZ/OX and the maximum non-linearity would be YZ/XZ.

Non-linearity is relatively unimportant in modern transducers, since they are usually interfaced to a computer system and the non-linearity can be removed by means of a 'look-up' table between the output and true displacement.

The *accuracy* of a transducer is the difference between the measured value (via the 'look-up' table in the above system) and the true value, and is usually expressed as a percentage of full scale.

The noise produced by a transducer, which limits its ability to detect a given signal, is known as its detectivity d, defined by

$$d = \frac{\text{signal-to-noise ratio at output}}{\text{input signal}}$$

The least detectable signal is defined as that input signal which produces an output r.m.s. signal-to-noise ratio of unity (in the chosen output bandwidth), so $d = 1/(\text{least detectable signal})$. The meaning of this is that if, in the absence of an input signal, the transducer produces a certain output noise power N, then the least input signal that can be detected is that which produces an output signal power S equal to N (that is, doubles the total output power). The term 'noise equivalent input signal' is sometimes used instead of 'least detectable signal'. In practice one finds that a signal can be detected with reasonable reliability if the output signal-to-noise ratio is unity though the 'least detectable signal' is strictly a definition. The units of d are in reciprocal input signal units; for example 10^6 m^{-1} for a displacement transducer that has a least detectable signal level of 10^{-6} m. It is important to notice that d automatically involves the *bandwidth* used in the system. Many transducers produce white noise, which has the same power at all frequencies, so the output noise power is proportional to the bandwidth used and *d* is inversely proportional to the square root of the bandwidth.

The *response time* of a transducer may be defined as the time taken to respond to 63 per cent (1–1/*e*) of its final value in response to a step (for a first-order system) or the time to reach its first peak (in an underdamped second-order system). It is related to the bandwidth of the transducer. Denoting response time by Δt and bandwidth by Δf, the relation $\Delta f \times \Delta t \approx 1$ holds approximately in most cases.

1.6 Measurement systems

A general measurement system is shown in figure 1.6. It is a more detailed version of figure 1.3, showing all the blocks in the measurement process.

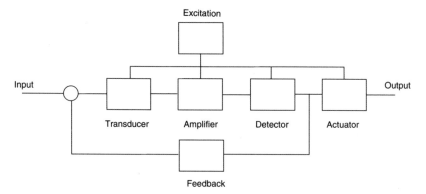

Figure 1.6 *A general measurement system.*

The transducer block, which may contain a sensor plus a bridge circuit or several transducers, produces the electrical output required by the amplifier, and usually requires some form of excitation. In simple systems, d.c. excitation is sufficient but most precision systems require a.c. excitation. In such cases the amplifier output will also be a.c. and a detector, usually a phase sensitive device (PSD), is used to convert the signal to d.c. for display or control purposes. A PSD requires a reference signal at the excitation frequency, and produces a d.c. output of polarity dependent on the phase of the input with respect to the reference; a.c. excitation avoids a serious type of electrical noise, known as 1/*f* noise, which becomes large at low frequencies and affects d.c. systems. An actuator provides the type of output required.

In some measurement systems, for example for measuring acceleration or force, it is possible to apply feedback from the output to the input, as shown in figure 1.6. This produces a much superior system with response determined by the (passive) elements in the feedback path, with improved linearity and insensitivity to changes in transducer responsivity and amplifier gain. It is not always possible to apply such feedback but it is always advantageous when it can be done.

Some measurement applications require only a few of the blocks in figure 1.6. A solar cell for measuring light levels could be represented by just the transducer block with no excitation. Similarly, a d.c. thermistor system for measuring temperature would require the transducer block (including a bridge circuit) and a d.c. amplifier. Such a system would be improved by using a.c. excitation, as explained above, and the PSD would then be added. Many systems, such as capacitive and inductive displacement measurement systems, are inherently a.c. and therefore require a.c. excitation and a PSD. A few systems, notably force feedback weighing machines, require all the blocks. Figure 1.7 shows such a system in which any imbalance of the beam is detected by the displacement transducer, amplified, converted to a d.c. voltage of appropriate polarity by the PSD, and finally applied to the actuator in the feedback loop to return the beam to balance. The force applied is directly proportional to the current flowing through the actuator coil, and absolute calibration is possible by simply finding the current required to balance a known mass m (producing force mg).

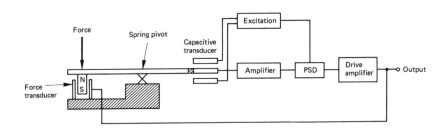

Figure 1.7   *Force–balance system.*

The transducer parameters discussed in section 1.5 also apply to complete measurement systems, of course. The system will have an overall responsivity determined by the product of transducer responsivity and gain (amplifier × PSD, etc.) in an open-loop system, or by the feedback components in a closed-loop system (for example, amps per newton for the system in figure 1.7). Similarly, the overall detectivity is determined mainly by the transducer and amplifier. With resistive transducers the amplifier noise can often be made negligible, but some transducers (such as capacitive) are inherently noiseless, and amplifier noise then determines the overall detectivity. In well-designed systems the PSD does not contribute to the noise.

1.7 Exercises

1.7.1. Explain the terms *responsivity, detectivity* and *range*, and give an example of each term by reference to a transducer of your choice for (a) displacement, (b) temperature and (c) light (visible or infra-red).

1.7.2. (a) State whether the following transducers are self-generators, modulators or modifiers:
    (i)   a rotary potentiometer for angle measurement
    (ii) a thermocouple
    (iii) a photoconductive cell
    (iv) a mercury-in-glass thermometer.
  (b) Give an example of each of the following (do not select those in part (a) above):
    (i)   an electrical–thermal–electrical modulator
    (ii) a mechanical–electrical self-generator
    (iii) a mechanical modifier
    (iv) a radiant self-generator
    (v) an electrical–mechanical–electrical modulator.

2 Analogies between Systems

2.1 Analogies

We saw in chapter 1 that transducers operate by transforming energy from one domain to another, such as mechanical to electrical in the case of a piezoelectric device. Interesting analogies exist between several of the basic types of energy or signal, and we will discuss them now in order to illuminate our later discussion and comparison of the types of transducer. It is well known, for example, that the flow of fluid through a pipe is analogous to that of current through a resistor. Consideration of such analogies is not only interesting and instructive in itself, but can have considerable practical application and can sometimes provide the insight required for the solution of a problem by transposing the problem into a more familiar domain. We will consider initially only mechanical and electrical systems, but later extend our view to all the types of energy and signal considered above. The reader is referred to the book by Shearer *et al.* (1971) for a comprehensive treatment of mechanical and electrical networks.

2.2 Mechanical and electrical systems

Figure 2.1 shows a simple mechanical system, comprising a mass M supported by a spring S; in practice some damping is always present and it is usual to indicate this schematically by the dashpot D (even when the only damping is air damping).

We will assume that the mass is constrained to move only vertically by means of frictionless rollers (not shown) and that a force f is applied by, say, a magnet attached to it and a coil fixed to the frame (again not shown). The velocity of the mass (and of the spring and dashpot) with respect to the frame is v.

The mass, spring and dashpot are the basic 'building blocks' of any mechanical system and are known as the *elements* of the system. There are only three such passive mechanical elements; note that two of them, the mass and spring, can store energy but the third, the dashpot, dissipates energy. The force f and the motion v are known as the *variables* of the system (we could choose displacement d or acceleration a, of course).

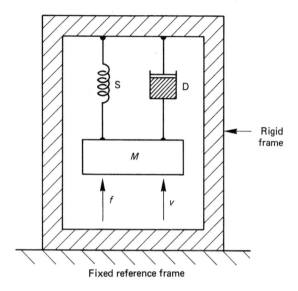

Figure 2.1 *Simple force-driven mechanical system.*

Each element in a mechanical system is defined by an equation relating it to the variables v and f:

mass M $\quad f = M\,\mathrm{d}v/\mathrm{d}t$

spring S $\quad f = K_\mathrm{m}d$ or $\mathrm{d}f/\mathrm{d}t = K_\mathrm{m}v$ (K_m is known as the spring stiffness)

Alternatively

$\quad\quad\quad d = C_\mathrm{m}f$ or $v = C_\mathrm{m}\,\mathrm{d}f/\mathrm{d}t$ (where $C_\mathrm{m} = 1/K_\mathrm{m}$ is the spring compliance)

dashpot D $\quad f = R_\mathrm{m}v$, assuming viscous damping (R_m is the viscous damping coefficient)

The motion of the mass is given by the differential equation

$$f = M\frac{\mathrm{d}v}{\mathrm{d}t} + R_\mathrm{m}V + \frac{1}{C_\mathrm{m}}\int V\mathrm{d}t$$

Its kinetic energy is $\frac{1}{2}Mv^2$, the potential energy of the spring $\frac{1}{2}C_\mathrm{m}f^2$ and the instantaneous power fv.

We will now consider a simple electrical system, developing similar equations with a view to deciding which (if any) of the variables and elements of the two systems can be considered to be analogous.

Figure 2.2 shows such a system, in which an e.m.f. e drives a current i through a series combination of an inductor, capacitor and resistor, producing a voltage V ($= e$) across the network.

Resistors, capacitors and inductors are the three basic passive building blocks in electrical systems, and are clearly the electrical elements; note that again there are two storage elements (capacitance and inductance) and one dissipative element (resistance). Similarly e.m.f. e (or voltage V) and current i (or charge q) are the variables.

The defining equations for the elements are:

resistance R: $V = iR$

or $i = GV$ (in terms of conductance $G = 1/R$)

capacitance C: $C = i\,dV/dt$

inductance L: $V = L\,di/dt$

The corresponding differential equation for the system is

$$V = L\frac{di}{dt} + Ri + \frac{1}{C}\int i\,dt$$

The energy in the inductor is $\frac{1}{2}Li^2$, in the capacitor $\frac{1}{2}CV_C^2$ and the instantaneous power is iV.

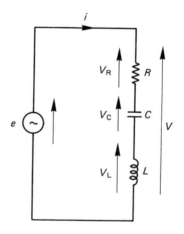

Figure 2.2 *Simple electrical system.*

For comparison purposes the equations have been grouped together in table 2.1, arranged arbitrarily at present.

Things are usually imperfect in the physical world (often the mental world is not much better) and this topic is no exception; there are apparently two solutions! If we consider that force f corresponds to voltage V and velocity v to current i, we find that the equations are comparable if mass M is analogous to inductance L, compliance C_m to capacitance C and viscous damping R_m to resistance R. Alternatively, we get an equally good match if we take force to correspond to current and velocity to correspond to voltage, giving mass analogous to capacitance, compliance to inductance and viscous resistance to electrical conductance.

Table 2.1 Defining equations of mechanical and electrical elements

	Storage	*Dissipative*
Mechanical mass M	spring S	dashpot D
$f = M \, dv/dt$	$df/dt = K_m v$ $v = C_m \, df/dt$	$f = R_m v$
$E = \frac{1}{2} M v^2$	$E = \frac{1}{2} C_m f^2$	$P = fv$
Electrical inductance L	capacitance C	resistance R conductance G
$V = L \, di/dt$	$i = C \, dV/dt$	$V = iR$ $i = GV$
$E = \frac{1}{2} L i^2$	$E = \frac{1}{2} C V^2$	$P = iV$

The form of the analogy thus depends on how we choose to compare the variables. The first choice above (force–voltage, velocity–current) is known as the force–flow analogy, comparing variables that physically have the effect of 'forcing' something to happen or 'flow' in a system. A force or e.m.f. can clearly be considered 'forcing' and a current or velocity (resulting from a force) are clearly 'flow' variables. Unfortunately this method of deciding on the type of variable is not foolproof; although one usually thinks of current flowing in response to an applied e.m.f. it is quite possible to apply a current generator to a circuit, 'forcing' a voltage to be produced across it. Similarly, a mechanical system may be 'velocity driven', if the complete system has a velocity impressed on it, as in an accelerometer where the frame is moved, producing a force on the suspended mass and a resulting motion.

The second choice above (force–current, velocity–voltage) is known as the through–across analogy, variables being compared in terms of whether they act 'through' or 'across' the system. There is no ambiguity here; strictly an across variable is one that has to be measured between two points in space (for example, voltage or displacement, since a displacement is always with respect to some frame of reference) and a through variable one that can be measured at one point in space (for example, current or force).

The two analogies are summarised in table 2.2. The force–flow analogy is more obvious physically, providing analogous elements that appear correct intuitively. However, it has the unfortunate consequence that, since it is not based on precise concepts, when using it one finds that elements in series in one domain (for example, mechanical) appear in parallel in the electrical analogue! This does not matter in simple systems, but for a complicated mechanical system it is far easier to use the through–across analogy, which leads to a direct one-to-one correspondence between elements. We will not pursue the matter here, though we will use an electrical analogue in analysing an accelerometer in chapter 8.

Table 2.2 Mechanical/electrical analogies

Analogy	Variables	Elements
force–flow	force–voltage velocity–current	mass–inductance compliance–capacitance resistance–resistance
through–across	force–current velocity–voltage	mass–capacitance compliance–inductance resistance–conductance

2.3 Fluid systems

A fluid system is of course mechanical, but we will consider such systems separately since they are rightly a class of their own.

Figure 2.3 shows a simple fluid system comprising a tank of water maintained at a constant height (by a tap and overflow) feeding a second tank via a narrow

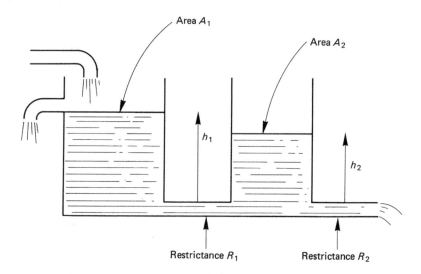

Figure 2.3 *Simple fluid system.*

tube. Water escapes from the second tank by another narrow tube.

The variables in a fluid system are pressure p and flow i (volume/s). The most obvious elements are fluid capacitance C_f and fluid restrictance R_f (or conductance G_f).

The basic equation for restrictance is $p = iR_f$, clearly analogous to Ohm's law, so we have the analogies pressure–voltage, flow–current and restrictance–resistance. In this case the force–flow and through–across classifications give the same result (it is only in the mechanical case above that problems arise). The volume V of a container of height h and cross-sectional area A is $V = Ah$ and comparing this with the equation $q = CV$ for electrical charge suggests the analogies volume–charge, area–capacitance and height–voltage. It is therefore easier to define restrictance by $h = iR_f$ and use height instead of pressure when dealing with simple tanks of fluid.

The most useful analogies are therefore height–voltage and flow–current for the variables, and cross-sectional area–capacitance and restrictance–resistance for the elements. There is an equivalent of electrical inductance, known as fluid inertance, arising because a fluid has mass which must be accelerated in moving it. However, in many cases it is small and can be neglected.

The analogies are summarised in table 2.3.

Table 2.3 Fluid/electrical analogies

Fluid	Inertance I $h = I \, di/dt$	Capacitance A $V = Ah$ $i = A \, dh/dt$	Resistance R_f $h = iR_f$
Electrical	Inductance L $v = L \, di/dt$	Capacitance C $q = Cv$ $i = C \, dv/dt$	Resistance R $v = iR$

2.4 Thermal systems

A simple thermal system is shown in figure 2.4, in which one end of a metal rod is maintained at a constant temperature in an oven and the other end attached to a large solid block at constant temperature.

The thermal elements are temperature T and heat flow i (watts). The equation for thermal conductance G_t is $i = G_t T$, equivalent to Ohm's law, so the analogies are heat flow–current, temperature (strictly temperature difference)–voltage and thermal conductance–conductance. Both variable classifications give the same result. The equation for thermal capacitance C_t is $H = C_t T$, where H is heat energy (as opposed to flow); this is directly analogous to $q = CV$ again, so we have heat–charge and thermal capacitance–capacitance. However, if we look for a property similar to inertia to find a second storage element, rather surprisingly there isn't

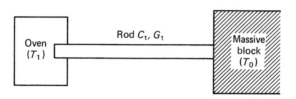

Figure 2.4 *Simple thermal system.*

one! Thermal energy is related to the motion of the electrons in the material and no 'acceleration' is required, making thermal equivalents very simple.

The analogies are summarised in table 2.4.

Table 2.4   Thermal/electrical analogues

Thermal	Capacitance $H = C_t T$ $i = C_t \, dV/dt$	Resistance/Conductance $i = G_t T$
Electrical	Capacitance $q = Cv$ $i = C \, dv/dt$	Resistance/Conductance $i = Gv$

2.5   Other systems: radiant, magnetic, chemical

Systems employing optical radiation are often essentially identical with the thermal systems discussed above. If the radiation is that from a hot body, there will be a stream of radiation emitted from the body and a detector intercepting it will rise in temperature according to its conductance and capacitance as in the equations above. The same applies to a beam of radiation from, say, a laser, though if it is detected by a photovoltaic or photoconductive detector (that is, a photon detector as opposed to a thermal detector) the analogy is no longer useful since the temperature rise of the detector is unimportant.

Distinct similarities exist between magnetic and electrical quantities, though the analogy is limited by the fact that magnetic monopoles apparently do not exist. They have long been sought experimentally, but without success. Magnetic flux ϕ is analogous to electrical current, being driven round a magnetic circuit of reluctance R_1 by a magnetomotive force M, according to the equation $M = R_1\phi$. However, there are no equivalents of electrical capacitance or inductance. The analogy is useful in analysing magnetic systems such as velocity transducers.

An analogy can be set up between chemical and electrical quantities, in terms of flow of ions and concentration gradients, but does not appear to have much practical application at present. However, chemical sensors have become increasingly important in recent years, and such devices are discussed in chapter 10.

There is an interesting theory that many non-physical systems can be analysed in terms of the three basic elements representing inertia, storage (or capacitance) and dissipation, and two variables representing force and flow. For example, factory assembly lines and mining involve these terms to different degrees. The concepts apply particularly well to the educational process in students, in which knowledge is the flowing variable and the lecturer the forcing variable. Students clearly display considerable inertia (in getting down to their studies), enormous dissipation (in forgetting taught facts) and almost zero capacity (for learning new ones)!

Table 2.5 summarises the variables, elements and equations for the systems discussed above. The variables have been chosen using the force–flow method and the elements aligned accordingly. Alternative variables and equations (such as current and charge) are shown where appropriate.

Table 2.5 Variables, elements and equations for physical systems

System	Variables Force	Variables Flow	Elements	Equations		
Electrical	v	i	L, C, R	$v = L\,di/dt$	$i = C\,dv/dt$	$v = Ri$
		q	G		$q = Cv$	$i = vG$
Mechanical (translational)	f	ν	M, C_m, R_m	$f = M\,d\nu/dt$	$\nu = C_m\,df/dt$	$f = R_m\nu$
		x			$x = C_m f$	
Fluid	$h(p)$	i	I, C_f, R_f	$h = I\,di/dt$	$i = C_f\,dh/dt$	$h = R_f i$
		V			$V = C_f h$	
Thermal (radiant)	T	i	C_t, R_t		$i = C_t\,dT/dt$	$T = R_t i$
		H	G_t		$h = C_t T$	$i = G_t T$
Magnetic	M	ϕ	R_1		$M = R_1\phi$	

2.6 Exercises

2.6.1. Explain how mechanical and electrical variables can be classified using the 'force–flow' and 'through–across' methods. Illustrate your answer by drawing the electrical analogues for figure 2.5, in which a mass–spring system is driven by a force f.

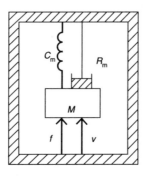

Figure 2.5

2.6.2. Draw the electrical equivalent of the simple thermal system of figure 2.4. Is this an accurate analogy?

2.6.3. (a) State a mechanical equivalent of an electrical transformer.
   (b) Can an electrical equivalent be drawn for all mechanical circuits?
   (c) Can a mechanical equivalent be drawn for all electrical circuits?
   (d) Do all thermal circuits have duals?
   (e) What is the inertial element in a liquid system?

3 Physical Effects available for Use in Transducers

3.1 Representation of transducers

An interesting three-dimensional representation of transducers was proposed by Middelhoek and Noorlag (1981b), shown in figure 3.1. The primary energy input to the system is represented by the x-axis and the energy output by the y-axis. Self-generating transducers therefore lie in the x–y plane. With the six forms of energy discussed above we have 36 possibilities, of which 30 are true self-generators and six modifiers (having the same form at both input and output). The most important self-generating input transducers are the five having electrical output, shown as crosses in figure 3.1; the most important actuators (having electrical input) are shown as squares, and the six modifiers as small circles.

The modulating transducers are represented by points in three-dimensional space, the z-component representing the modulating (signal) input. There are evidently 216 modulators in all; the most important are those for which both input and output energy are electrical, of which there are five (shown by dots in the figure), though there are few known devices with non-electrical x- and y-components.

There are, of course, a lot of gaps in figure 3.1. However, what is particularly interesting is that the figure apparently can accommodate all known (or possible) transducers, and as new ones are developed the gaps can simply be filled in. It is instructive, though often very difficult or even impossible, to pick a point and try to think of a transducer that could occupy it.

Table 3.1 summarises the most important transducers currently in use. However, since chemical energy is very distinct from the other forms (which can all be described as 'physical') and since chemical measurement is a large and important subject in its own right, we have omitted chemical transducers and will not consider them further except for some discussion of recently developed solid-state devices in chapter 10.

We will consider the basic physical processes available for use in transducers in the remainder of this chapter, grouping the processes in terms of transducer type, that is, self-generators, modulators and modifiers. A few applications will be mentioned for completeness, but we will concentrate on the physical principles involved and leave the detailed discussions of specific devices to later chapters.

Table 3.1 The most important physical effects and associated transducers

Type	Transduction	Physical effect	Application
Self-generators	radiant–electrical	photovoltaic; radiation–current	solar cells
	mechanical–electrical	electrodynamic; velocity–voltage	tachogenerators
		piezoelectric; deformation–charge	piezotransducers
	thermal–electrical	thermoelectric; temperature–voltage	thermocouples
		pyroelectric; temperature–charge	radiation detectors
	magnetic–electrical	electromagnetic; flux change–voltage	magnetic fields
Modulators	electrical–(radiant)–electrical	photoconductive; radiation–resistance change	radiation detectors
		photoemissive; radiation–current	
	electrical–(mechanical)–electrical	piezoresistive; strain–resistance change	strain gauges
		displacement–impedance change	electrical displacement transducers
	electrical–(thermal)–electrical	thermoresistive; temperature–resistance change	thermistors, resistance thermometers
	electrical–(magnetic)–electrical	magnetoresistive; magnetic field–resistance change	magnetic field measurement
		Hall effect; e.m.f. with applied current in magnetic field	Hall effect probes for current or fields
	radiant–(mechanical)–radiant	radiation change due to motion	encoders and gratings
Modifiers	radiant–radiant	temperature change due to collected radiation	thermal radiation detectors
	mechanical–electrical	displacement change due to pressure	diaphragm pressure transducers
		displacement change due to force	force transducers
		pressure change due to flow	orifice-type flow transducers
	thermal–thermal	temperature change due to heat flow	heat flux detectors
	electrical–electrical	change in electrical form	amplifiers, filters, etc.

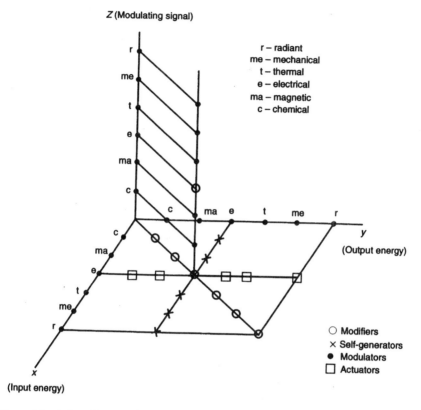

Figure 3.1 *A three-dimensional representation of transducers.*

3.2 Self-generators

We will consider here the four most important self-generators (having electrical output). Most self-generators are reversible, becoming actuators, and the important actuators will also be discussed.

3.2.1 *Radiant self-generators: the photovoltaic effect*

Radiant self-generators are of considerable interest, in addition to the transducer field, because of the importance of converting radiant energy to electrical. Photovoltaic transducers are in fact none other than the well-known silicon solar cells used for satellite power supplies. They are essentially semiconductor diodes in which light is permitted to fall on the junction region and are indistinguishable from diodes in the dark (apart from a high reverse current).

When a junction diode is produced, the positive charge carriers in the p material tend to flow to the n material, and vice versa for the n material. The p material becomes negatively charged and the n material positively charged, so an electric field (the junction field) is developed in the junction region to stabilise the flow of charge, directed as shown in figure 3.2.

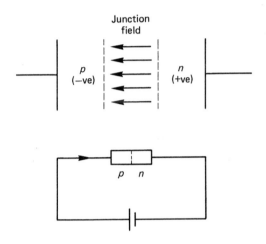

Figure 3.2 p–n *diode.*

Forward bias reduces the field and the current increases exponentially. Reverse bias increases the field and reduces the current, though a small leakage current remains.

When light falls on a photoconductive material electrons may be excited across the energy gap E_g, provided that the quantum energy $h\nu$ is greater than E_g, so that the conductivity is increased. The electron in the conduction band and the resulting hole in the valance band are no longer tied together and are therefore free to move. In a p–n junction device the free electron–hole pair comes under the influence of the junction field (provided that the photon is absorbed in the junction region) and the hole is swept to the left (to the p material), and the electron to the right. The hole and electron are thus physically separated and may flow in an external circuit. Moreover, the direction of current flow is in the reverse current direction (since the holes go to the left), so the resulting light current I_L is seen as a large increase in the reverse current.

The responsivity is easily evaluated. If we have an incoming stream of monochromatic radiation W of frequency ν, the number of photons/s is $W/h\nu$ and the resulting light current is $I_L = qeW/h\nu$ where q is an efficiency factor. The responsivity r is thus $qe/h\nu$ A/W, and has a value of about 0.1 A/W at 1 μm. Devices are produced by depositing a layer of say p material on a suitable substrate and following this by a thin layer of *n* material, producing an extended junction. They are usually circular or rectangular and may vary in area between about 1 mm^2 and 10 cm^2.

The reverse of the photovoltaic effect, in which light is emitted from a forward-biased diode when current is passed through it, does occur but unfortunately not with the same device. Electron–hole pairs are created using energy from the electric field applied and their recombination produces light (under suitable conditions). This occurs with gallium arsenide and gallium phosphide, but with silicon and germanium (the most widely used photovoltaic devices) most of the energy is dissipated as heat. Light-emitting diodes (LEDs) such as gallium arsenide produce incoherent light, but laser action can be produced by depositing parallel reflecting surfaces on the semiconductor crystal.

An LED can be persuaded to show a very small photovoltaic effect, producing a small current if very bright light (for example, from a laser) is shone on it, but a silicon solar cell will not emit any light however much current is applied.

3.2.2 *Mechanical self-generators: the piezoelectric effect*

There are two mechanical self-generators, employing the piezoelectric effect and the electrodynamic effect. However, the latter is closely associated with the electromagnetic effect and will be considered under magnetic self-generators in section 3.2.4.

Piezoelectric effect

The word 'piezo' means 'push' and the effect is exhibited by the appearance of a current or voltage when a force is applied to a suitable material. The name is slightly misleading because physically it is the dimensional change due to the force that is important, producing a surface charge density on the faces of the device. It is strictly a displacement-to-charge converter and is somewhat unusual in that most self-generators transduce between analogous variables (on the through–across classification), whereas here the transduction is between displacement (across) and charge (through). The effect does not occur in materials having a symmetrical charge distribution, since clearly there is no reason for one surface to be favoured rather than another, but occurs only in crystals of certain types. The best known naturally occurring material is quartz but the effect can be produced artificially in ferroelectric materials (such as lead zirconate titanate) by heating them in a strong electric field. The magnitude and direction of the effect depend on the direction in which the crystal is cut with respect to its lattice, and tables of the relative coefficients are available.

For a small disc of quartz, as in figure 3.3, the charge density q is given in terms of the applied force f by

$$q = df \qquad (3.1)$$

where d is known as the 'd coefficient' for want of a better name. d is typically 2×10^{-12} C/N for quartz and about 150×10^{-12} C/N for PbZ. The opposite surfaces of the device are metallised, producing a capacitor of capacitance $C = \epsilon\epsilon_0 A/t$ where ϵ is the relative permittivity (about 4.5 for quartz, 1800 for PbZ), ϵ_0 the permittivity for free space ($10^{-9}/36\pi$ F/m), A the area and t the thickness. With $A = 1$ cm^2 and $t = 1$ mm we find $C = 4$ pF (1600 pF for PbZ). The voltage corresponding to the charge can be found, and this is often given in terms of the 'g coefficient' in the relationship

$$V = gtP \qquad (3.2)$$

where P is the pressure applied. g ($= d/\epsilon\epsilon_0$) has a value of about 5×10^{-12} C/N for quartz and about 10^{-2} V m/N for PbZ.

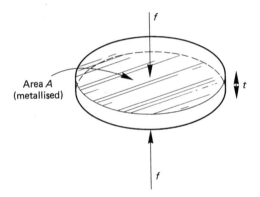

Figure 3.3　*Piezoelectric transducer.*

Piezoelectric devices are often used for force and pressure measurement, but it is important to note that their response to displacement does not extend to d.c. because of series capacitance. They are also widely used in accelerometers, a small crystal performing the functions both of supporting and detection of the relative displacement of the mass.

As with most self-generators the effect is reversible, and the application of a voltage V to a piezo crystal (producing a charge via the crystal capacitance) leads to a small displacement x, given by $x = dV$. d is the same constant as in the formula $q = df$ above. The effect has been used in ruling engines for producing diffraction gratings and in stabilising the cavity length in lasers, but is perhaps better known for the hourly or more frequent tones emitted by digital watches!

3.2.3 Thermal self-generators: the thermoelectric and pyroelectric effects

Thermoelectric effect

This is another name for the Seebeck effect, whereby an e.m.f. occurs in a circuit comprising two different metals if the junctions between them are at different temperatures, as in figure 3.4. An e.m.f. $e = P\delta T$ is observed, P being known as the thermoelectric power, which may be positive or negative, the resulting effect being proportional to the difference. The magnitudes are fairly small, being of the order 10–100 $\mu V/°C$.

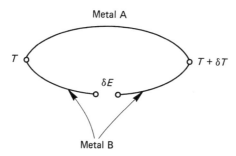

Figure 3.4 *Basic thermocouple.*

The effect is due to the equalisation of Fermi levels when two metals are placed in contact. For each metal the energy levels are filled up to a certain value known as the Fermi level, and the levels rapidly become equal when the contact is made, the resulting e.m.f. being the difference between the two levels.

The Seebeck effect is reversible, and the Peltier effect is the heating or cooling of a junction when a current flows in the circuit. A further reversible effect is the Thomson effect, which is related to the temperature gradient in the conductors between the junctions, and leads to additional heat flow or voltage. The Thomson effect is relatively small, but leads to a second-order term in the simple equation above for the Seebeck effect.

There are several laws relating to thermocouples, such as the laws of intermediate metals and temperatures, whose lengthy formal statements appear to have delighted some authors in the past, though probably not their readers since these laws are all intuitively obvious.

Pyroelectric effect

This is the thermal equivalent of the piezoelectric effect, in which deformation produces a surface charge density. The word means 'furnace electricity' and a temperature difference across a disc of pyroelectric material thus produces a charge density. Most piezoelectric materials show the effect, especially semiconductor devices, and it is a serious disadvantage in some cases.

The main practical application is in the measurement of infra-red radiation, typically in intruder detection. The stream of radiation is directed on to a small disc of material whose faces are metallised, and a corresponding voltage is produced. The best-known material is lead zirconate titanate. As with piezoelectric devices, the response does not extend to d.c. and special techniques are required for intruder detection applications.

3.2.4 Magnetic self-generators: the electromagnetic and electrodynamic effects

Faraday's law of induction states that the e.m.f. produced in a coil of n turns due to a changing magnetic flux ϕ is

$$e = -n \, d\phi/dt$$

The effect can be used directly for the measurement of changing magnetic fields or of steady fields by rotating the coil at a known rate, when it is known as the *electromagnetic effect*. However, Faraday's law is most useful for measuring the velocity of a conductor moving in a magnetic field, in which case the transduction action is strictly that of a mechanical self-generator, though it will be covered here for completeness. It is often known as the *electrodynamic effect*.

By considering a straight section of conductor of length l moving with velocity v perpendicular to a magnetic field of induction B, as in figure 3.5, it is easily deduced that an e.m.f. is produced given by $e = (Bl)v$. The same formula applies if the conductor is a coil of total length l.

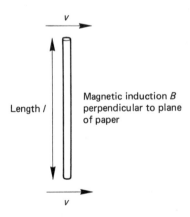

Magnetic induction B perpendicular to plane of paper

Length l

Figure 3.5 *E.m.f. in a conductor moving in a magnetic field.*

The effect is reversible, so a current i fed to a coil produces a force $F = (Bl)i$ newtons. The identity of the coefficients (Bl) in the two cases is very useful in calibrating instruments that use a magnet/coil system for velocity measurement, since a known force can easily be applied by simply adding a small mass.

Most practical devices consist of a fixed magnet with a movable coil attached to the object whose velocity is required, though the reverse configuration is sometimes used. Rotational devices, usually moving coil, are also very widely used, usually being referred to as tachogenerators.

A further application of Faraday's law is in magnetoresistive and Hall-effect transducers, in which the motion of charge carriers in a magnetic field produces a resistance change or e.m.f. These effects are discussed in section 3.3.4 which deals with magnetic modulators.

3.3 Modulators

This is the largest group of transducers. We will consider specifically the five modulators having both electrical input and electrical output, along with a few more general modulators with non-electrical input energy.

3.3.1 Radiant modulators: the photoelectric and photoconductive effects

Photoelectric effect

The photoelectric effect is the emission of electrons from a metal surface when light of a suitable wavelength falls on it. Such a surface is characterised by a work function ϕ, which is the amount of energy required to withdraw an electron from it (that is, to infinity). If the light is monochromatic, of frequency ν and wavelength λ, the condition for emission of an electron is

$$h\nu \geq \phi$$

where h is Planck's constant. Transducers using the photoelectric effect are known as photoemissive detectors and consist of a cathode of suitable material and an anode at a potential of, say, 100 V enclosed in an evacuated jacket. All the electrons emitted are collected by the anode, so a steady photocurrent flows in response to a steady illumination. For an incident radiation of W watts, the number of photons per second is $W/h\nu$ and the photocurrent is $eqW/h\nu$, where q is an efficiency factor, so the responsivity is

$$r = eq/h\nu \text{ A/W}$$

The responsivity thus increases with wavelength, reaching a maximum for $h\nu = \phi$, after which it rapidly falls to zero. In practice the maximum wavelength is about 1 μm, for a cathode material of caesium oxide and silver.

Photoemissive detectors are not much used now, but the same effect is employed in photomultipliers, in which the photocurrent is amplified by a series of secondary electrodes, producing very high responsivity and detectivity.

Photoconductive effect

Photoconductive materials are semiconductors in which a transition between valence and conduction bands, separated by an energy gap E_g, may be excited by an incident photon of suitable wavelength. Unlike the photovoltaic effect, in which a physical separation of charge carriers occurs, there is simply a change in conductivity as the name implies. As above, the condition for excitation is $h\nu \geq E_g$.

The conductivity of a semiconductor is given by $\sigma = Neu$, where N is the total number of electrons in the conduction band, e the electronic charge and u the mobility of the charge carriers. N is strongly dependent on temperature, according to the formula

$$N = N_0 \exp(-E_g/2kT) \qquad (3.3)$$

where N_0 is the total number of electrons in the material (that is, in the valence and conduction bands) and k is Boltzmann's constant. N therefore increases with temperature and is zero at absolute zero.

If the incident radiant power is W, the number of carriers produced per second is $qW/h\nu$, where q is an efficiency factor. Unlike the photovoltaic effect, the carriers have a limited lifetime τ in the conduction band, and it is easy to show that the steady-state excess carriers δN due to the incident radiation W is equal to $qW\tau/h\nu$. The fractional change in conductivity is the same as the fractional change in bulk resistance, $\delta R/R$, and is

$$\frac{\delta\sigma}{\sigma} = \frac{\delta R}{R} = \frac{qW\tau}{h\nu N}$$

It is clear that we require a long lifetime τ to get a large response, though clearly this will limit the frequency response of the device to changing incident light. Also, we require the number of electrons N in the conduction band to be small, so that the element's resistance will be high. The responsivity will increase with λ, as for the photoemissive device, falling rapidly to zero once the condition $h\nu = E_g$ is reached (that is, for $\lambda > hc/E_g$ where c is the velocity of light). One of the best-known photoconductors is lead sulphide, which responds out to about 3 μm.

In order to have a response far into the infra-red we require E_g to be small; unfortunately this means that at, say, room temperature N will be large (since a lot of electrons will be excited thermally) so $\delta R/R$ will be small. It appears that we cannot get both a large resistance change and a response far into the infra-red at the same time. Rather surprisingly, since munificence is not often exhibited in physics (or by physicists for that matter), it is possible to obtain both features together by simply cooling the detector. This is usually done with liquid nitrogen (77K) and greatly reduces the thermal excitation so N becomes small and R large. This is done, for example, with detectors employing indium antimonide, which has the smallest energy gap of any undoped material, and responds out to about 6 μm. It is possible to obtain a response at larger wavelengths by doping the material to produce additional energy levels within the normal energy gap; such devices are known as extrinsic photoconductors, in contrast to the intrinsic types described above.

A further class of photoconductors employs what is known as the charge amplification effect. In some materials, notably cadmium sulphide, a hole-trapping effect occurs owing to impurities (copper ions). The effective lifetime of the carriers is greatly increased and the devices have very high responsivity though very low frequency response. Cadmium sulphide has a large energy gap, with peak response about 0.6 μm, somewhat similar to that of the eye.

3.3.2 *Mechanical modulators*

Strain gauges (and the piezoresistive effect)

Strain gauges employ the change in resistance of a suitable material when subjected to an applied stress, for the measurement of displacement or strain. The simplest form of device is a cylinder of area A, diameter d and length l of a material of resistivity ρ, subject to a force f, as shown in figure 3.6.

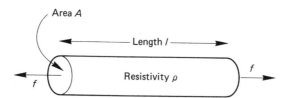

Figure 3.6 *Strain gauge transducer.*

The bulk resistance $R = \rho l / A$ so differentiating logarithmically

$$\frac{\delta R}{R} = \frac{\delta l}{l} - \frac{\delta A}{A} + \frac{\delta \rho}{\rho}$$

For a cylinder, $A = \pi d^2/4$, so $\delta A/A = 2\delta d/d$. The relationship between the change in diameter and change in length is given by Poisson's ratio $\nu = -(\delta d/d)/(\delta l/l)$: for a homogeneous material $\nu = 0.5$. Thus

$$\frac{\delta R}{R} = \frac{\delta l}{l} + \frac{2v\delta l}{l} + \frac{\delta \rho}{\rho}$$

The gauge factor (GF) is defined as the ratio of fractional change in resistance to fractional change in strain

$$\mathrm{GF} = \frac{\delta R/R}{\delta l/l} = 1 + 2v + \frac{\delta \rho/\rho}{\delta l/l}$$

The second term is entirely due to dimensional changes, whereas the third is known as the piezoresistive term. The piezoresistive effect is the change in actual resistivity due to applied strain; it is zero in metals but may be large in some semiconductors. There are thus two distinct types of strain gauge: metallic types, with GF about 2 since v is close to 0.5, and semiconductor types with GF about 100. Unfortunately, semiconductor devices also have a large temperature coefficient (being similar to thermistors) so that special temperature-compensation techniques must be used. Four devices are often used in a bridge arrangement, two being positioned in places of zero strain, to balance the effects of temperature. However, semiconductor devices are finding increased application, since a film of the material can often be deposited as an integral part of some other transducer – for example, in pressure or force measurement.

Electrical displacement transducers

These form a large and important class of devices in which a mechanical displacement changes the value of one of the electrical elements L, C or R, a bridge circuit being used to detect the change.

Resistive displacement transducers are simply rotary or linear potentiometers, as shown in figure 3.7. They are widely used but suffer from problems of wear, friction and limited resolution.

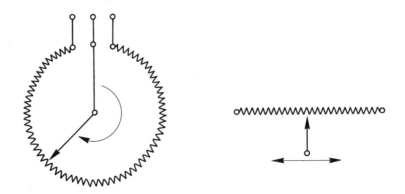

Figure 3.7 *Resistive displacement transducers.*

The capacitance C of a parallel plate capacitor of area A and plate separation d and containing a dielectric of relative permittivity ϵ is given by

$$C = \frac{\epsilon \epsilon_0 A}{d}$$

where ϵ_0 is the permittivity of free space ($10^{-9}/36\pi$ F/m). It is clear that the capacitance may be modified by changing ϵ, A or d, giving rise to the three basic types of capacitive displacement transducer: variable permittivity, variable area and variable separation. Typical devices are shown in figure 3.8; three-plate transducers are usually used in practice.

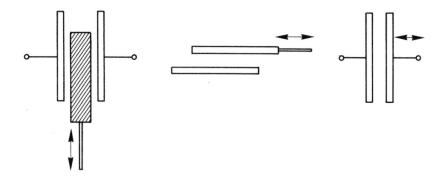

Figure 3.8 *Capacitive displacement transducers.*

The variable-permittivity capacitance transducer is little used; however the equivalent arrangement in inductive transducers is the only method used. The inductance of a toroid of relative permeability μ and area A, containing a coil of n turns and total wire length l, is given by

$$L = \frac{\mu_0 \mu n^2 A}{l}$$

where μ_0 is the permeability of free space ($4\pi \times 10^{-7}$ H/m). Such a device is not usable as a displacement transducer, and an 'opened out' version in the form of a cylinder and coil, as shown in figure 3.9, has to be used. Unfortunately the inductance is not easily calculable since the flux is no longer restricted to the magnetic material; the same formula applies but the effective μ is greatly reduced.

Referring to the above formula, the only way the device can be used to measure displacement is to vary the effective μ, since n, A and l cannot easily be changed. There are three important types: variable coupling in which the relative inductance of two coils is varied, the differential transformer in which the

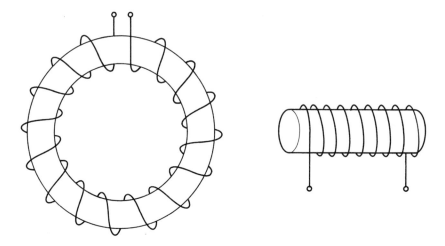

Figure 3.9 *Toroidal and cylindrical inductors.*

coupling between a primary winding and two secondaries is changed by the core, and variable reluctance in which the reluctance of a magnetic circuit is changed by a thin magnetic disc.

3.3.3 Thermal modulators

Thermoresistive transducers

Most thermal modulators are devices whose resistance changes in response to temperature. Both metallic and semiconductor sensors are widely used, though their characteristics differ greatly. The energy level diagrams for a metal and a semiconductor are shown in figure 3.10.

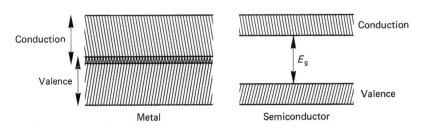

Figure 3.10 *Conduction and valence bands in metals and semiconductors.*

In metals the valence and conduction bands overlap, so there are always some conduction electrons and the material has substantial conductivity. In semiconductors there is a gap between the two bands, and the number of electrons N in the conduction band for a given gap depends on temperature T, according to equation (3.3) above.

The resistivity of most materials can be written as

$$\rho_T = \rho_\infty \exp(\beta/T)$$

where ρ_T is the resistivity at temperature T, ρ_∞ that at very high temperature and β a constant proportional to the energy gap. Considering two temperatures, T_1 and T_0, we can eliminate ρ_∞ and write the bulk resistance R as

$$R_{T_1} = R_{T_0} \exp\left[\beta\left(\frac{1}{T_1} - \frac{1}{T_0}\right)\right] \tag{3.4}$$

For semiconductors β is relatively large and positive, typically several thousand. The change of resistance with temperature is inherently exponential which is a disadvantage, though techniques exist for linearising the response. Most devices are in the shape of beads, bars or discs and are composed of oxides of nickel, cobalt or manganese. They are known as negative temperature-coefficient (NTC) thermistors, since the slope of the curve is negative. It is possible to obtain a positive slope (PTC devices) over a limited temperature range by suitable doping, but the exact response varies considerably from device to device.

In the case of metals, where an overlap occurs between the valence and conduction bands, the constant β can be considered to be small and negative so the exponential in equation (3.4) can be expanded. The resistance temperature coefficient a_T is given by

$$a_T = \frac{1}{R}\frac{dR}{dT}$$

and is approximately constant, so we obtain the familiar formula for resistance as a function of temperature, that is

$$R_T \approx R_{T_0}\lfloor 1 + a(T_1 - T_0)\rfloor$$

The curve has a small positive slope, depending on the particular metal. Figure 3.11 shows the change in relative resistance with temperature for both metals and semiconductors.

p--n *junction devices*

It is found that when a silicon diode is forward-biased and carrying a constant

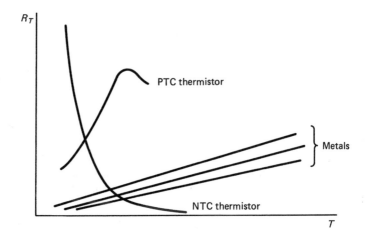

Figure 3.11 *Variation of resistance with temperature for metals and thermistors.*

current, the temperature coefficient of the voltage drop across it is approximately -2 mV/°C. The exact value varies between individual devices, so calibration is necessary, but the relation is essentially linear (unlike thermistors) and the response time is short. Such devices are particularly cheap, of course.

3.3.4   Magnetic modulators: the magnetoresistive and Hall effects

The magnetoresistive effect is the change in resistance of a semiconductor material subject to a magnetic field. It is closely associated with the Hall effect, which will be discussed first, although it is strictly a magnetic–electrical–electrical modulator.

The laws of electromagnetic induction discussed above for magnetic–electrical self-generators and tachogenerators also apply to the movement of individual charge carriers in a magnetic field. When a flat conductor carrying a current i is placed in a magnetic field of induction B normal to its surface, as in figure 3.12, an e.m.f. e is produced across the width of the conductor.

For a conductor of thickness t the e.m.f. is given by $e = K_H Bi/t$, where K_H is the Hall coefficient, dependent on the product of the charge mobility and resistivity of the conductor. The effect is negligibly small in most metals (which have low resistivity) and insulators (which have low mobility) but is appreciable in some semiconductors. Silicon and germanium can be used, but have fairly high resistivity, but indium antimonide is widely used, having $K_H \approx 20$ V/T.

Hall-effect transducers may be used for the non-contact measurement of current. A Hall probe (excited at constant current) is placed close to the wire carrying the unknown current and the magnetic field due to this current leads to

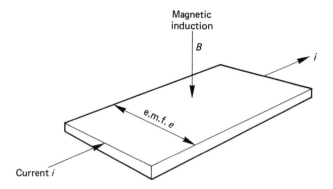

Figure 3.12 *The Hall effect.*

the Hall e.m.f. The device therefore acts as an electrical–magnetic–electrical modulator. They may also be used for the direct measurement of magnetic field, of course. The e.m.f.s produced are only a few microvolts and the effect is strongly temperature-dependent, so the detectivities obtained are low. The related magnetoresistive effect arises when the Hall voltage is short-circuited; the charge carriers are then deflected, resulting in an increased path length and consequently increased resistance. This effect is not much used at present, but its importance will increase as solid-state sensors are developed.

3.4 Modifiers

Since there are six basic forms of energy there should be six modifiers, having the same form of energy at input and output. Most modifiers convert energy between the two variables in the system – for example, force–velocity, flow–pressure or radiation–temperature. However, the only large groups of devices are the mechanical and radiant (or thermal) modifiers.

3.4.1 Radiant modifiers

Radiant modifiers are used in thermal radiation detectors in which an input stream of radiation falls on a blackened collecting screen, producing a rise in temperature which is detected by a thermocouple or other temperature transducer. The process is exactly analogous to the charging of a capacitor by a current, as shown in figure 3.13.

The temperature rise is given by

$$\delta T = \frac{\epsilon W / G}{(1 + j\omega\tau)}$$

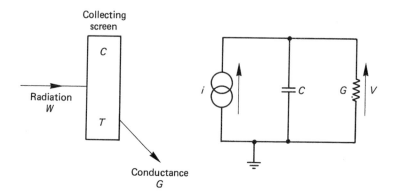

Figure 3.13 *Radiant modifier and its electrical equivalent.*

where ϵ is the emissivity of the screen and $\tau \, (= C/G)$ the thermal constant.

A rather similar form of thermal modulator is used in anemometers, in which a flow of fluid changes the temperature of a heated wire, transducing heat flow into temperature rise.

3.4.2 Mechanical modifiers

Mechanical modifiers convert a mechanical input into another mechanical form. One large class of devices is elastic elements, usually used for force measurement, in which the force is applied to a spring, cantilever beam or bar, producing a deflection (which is measured by a second transducer). The dimensions of the device are chosen to suit the range of forces of interest, and the displacement per unit force is easily calculable from the appropriate modulus of the material used. It is often convenient to measure strain rather than deflection, which is calculated from the well-known equations for a stressed bar.

Another important class of modifiers is pressure-sensitive components, such as tubes and bellows. Much ingenuity has been displayed by engineers in designing hollow tubes of various shape (C-type, spiral, helical, etc.), but such devices are rather crude in principle and often unsatisfactory in operation. A stretched diaphragm or thin plate that is deflected by the applied pressure is more satisfactory, the displacement being conveniently detected by capacitive or inductive transducers. These devices are linear over only a small range of deflections (typically about half the thickness of the diaphragm), and corrugated diaphragms are sometimes used to give a greater range.

A third class of mechanical modifiers is used for flow measurement. The principle mostly employed is that of providing some restriction to the flow and measuring the corresponding pressure drop, in analogy to measuring an electric current by finding the voltage across a small series resistor. Such devices are not very satisfactory, since the flow varies across the tube and may be streamline (without eddies, etc.) or turbulent depending on the appropriate Reynolds number (a function of velocity, diameter, density and viscosity); in addition, the flow rate is usually proportional to the square root of the pressure drop.

In recent years there has been much interest in resonant sensor systems, in which the resonant frequency of a mechanical structure varies with some parameter such as pressure or force. Such devices are also mechanical modifiers, and will be discussed in chapter 10, which deals with recent developments and future trends.

3.5 Exercises

3.5.1. (a) Explain how transducers can be classified on the basis of energy conversion into modifiers, modulators and self-generators, explaining the essential difference between each type.

      (b) List the most important types of self-generators (which produce an electrical output), the most important modulators (having electrical excitation and providing an electrical output), and the principal modifiers, in each case stating the physical principle involved and giving an example of a transducer employing that principle.

3.5.2. An alternative method of classifying transducers is into 'transforming transducers', in which the action is a transduction between corresponding variables (through–across) and 'gyrating transducers', in which a 'through' variable is transduced into an 'across' variable and vice versa. Classify the following transducers on this basis:

      (a) the main types of self-generator (photovoltaic, pyroelectric, electromagnetic, piezoelectric, thermoelectric)

      (b) the main types of modifier (mechanical, radiant, thermal).

3.5.3. Discuss the various applications of Faraday's law of induction to transducers, stating in each case the precise physical effect and the type of transducer involved (such as magnetic–electrical self-generator, electrical–magnetic–electrical modulator, etc.).

4 Transducer Bridges and Amplifiers

In the previous chapter we discussed transducers for all the main forms of energy except the electrical form. Electrical transducers may be modifiers, self-generators or modulators, of course, but the most important here are those having an electrical output, such as amplifiers and bridge circuits, since these are usually required to follow the transducers already discussed. Both are strictly modulators, since they involve a separate power source which is modulated by the (small) electrical signal. It is necessary to consider these devices at this stage so that we can deduce the responsivities of the transducers to be discussed in the next chapter.

4.1 Transducer bridges

As we have seen above, many transducers are modulators and control the flow of (usually electrical) energy. For example, a thermistor transduces temperature into change in resistance, but to obtain a usable signal a current must be applied and the resulting voltage measured. In general, where the transduction is into an electrical form, such as resistance, capacitance or inductance, some form of bridge circuit is required. Bridges may be excited by direct current, in which case they are always resistive, or by alternating current, when they may be resistive, capacitive or inductive.

4.1.1 D.c. bridges (resistive)

The most general form of resistive bridge is the four-arm arrangement, shown in figure 4.1 together with its equivalent circuit. Any one or all of the resistors may actually represent transducers. In measuring temperature one arm may be a thermistor and the others fixed (or variable) resistors; alternatively, in measuring strain with a cantilever beam, all four arms could be suitably connected strain gauges.

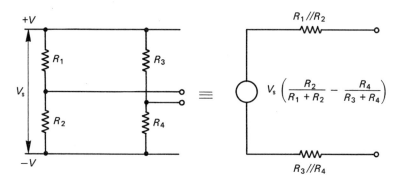

Figure 4.1 *General d.c. resistive bridge and its equivalent circuit.*

There are three cases of interest.

(i)   Double push–pull: all four arms are identical transducers, with
$$R_1 = R_2 = R_3 = R_4 = R$$
and
$$\delta R_1 = -\delta R_2 = -\delta R_3 = \delta R_4 = \delta R$$

The equivalent circuit reduces to that in figure 4.2(a).

(ii)   Single push–pull: R_1 and R_2 are transducers with
$$R_1 = R_2 = R$$
and
$$\delta R_1 = -\delta R_2 = \delta R$$

R_3 and R_4 are fixed and equal (often equal to R).
The equivalent circuit becomes that in figure 4.2(b).

(iii) Simple transducer: R_1 is the only transducer. Usually
$$R_1 = R_2 = R_3 = R_4 = R$$

We now have the equivalent circuit of figure 4.2(c).

Note that in figure 4.2(c) the resistance of R_1 in parallel with R_2 does not remain constant, unlike cases (a) and (b) where $\delta R_1 = -\delta R_2$. This causes a non-linear output for large changes in R_1.

4.1.2   A.c. bridges

An a.c.-excited resistive bridge could, of course, be obtained by simply replacing

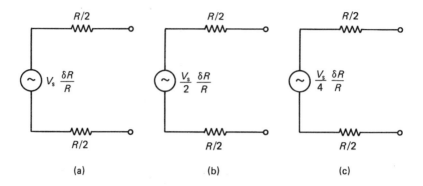

Figure 4.2 *Equivalent circuits of d.c. bridge arrangements.*

the d.c. excitation of figure 4.1 by a simple oscillator. However, it is generally preferable to use a transformer, as shown in figure 4.3, since this gives better isolation between the oscillator and the bridge, but in particular permits the two sides of the bridge to be exactly equal and opposite in excitation (or any other ratio that one wishes). It is very easy to wind a transformer with identical secondaries, because one simply uses two pieces of wire twisted together. Another advantage is that the bridge requires only two resistors (both of which are usually transducers) and a single-ended output is easily obtained by grounding the centre part of the secondary.

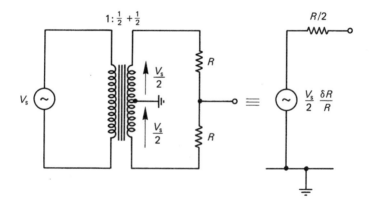

Figure 4.3 *A.c. resistive bridge and its equivalent circuit.*

Inductive and capacitive bridges are obtained in the same way. Push–pull types are preferred and the arrangements and equivalent circuits are shown in figure 4.4(a) and (b).

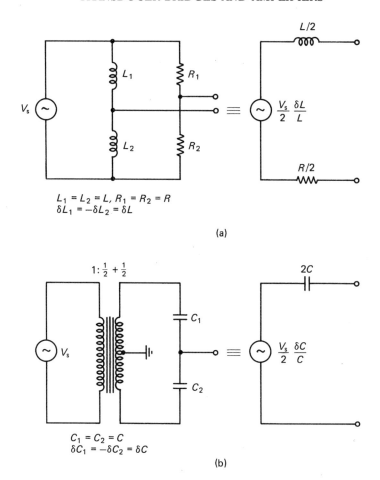

$L_1 = L_2 = L, R_1 = R_2 = R$
$\delta L_1 = -\delta L_2 = \delta L$

(a)

$C_1 = C_2 = C$
$\delta C_1 = -\delta C_2 = \delta C$

(b)

Figure 4.4 *Capacitive and inductive bridges and their equivalent circuits.*

4.2 Transducer amplifiers

We will summarise here the arrangement and properties of the most useful forms of operational amplifier (op amp) configurations for use with transducers. The basic amplifier is represented by a triangle as shown in figure 4.5. It is a differential device of (ideally) infinite gain, wide bandwidth, very high input impedance and negligible output impedance, the open-loop output voltage being given by

$$V_{\text{out}} = A(V_{\text{in}} + - V_{\text{in}} -)$$

where the open-loop gain A is very high. The application of negative feedback leads to a number of useful arrangements of stable and precisely known response. In each case the feedback tends to make the voltage difference between the two inputs of the op amp very close to zero.

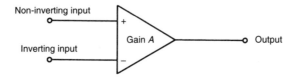

Figure 4.5 *Ideal operational amplifier.*

(i) Inverting amplifier (virtual-earth amplifier)

One of the most familiar arrangements is the virtual-earth amplifier of figure 4.6(a). Since the non-inverting input is grounded, the output will be such that the inverting input is also held close to ground by the feedback. It is easy to see that the voltage gain is $-R_2/R_1$ and that the input impedance is R_1. This circuit has the disadvantage that the input impedance may be low if high gains are required.

(ii) Non-inverting amplifier

This is not a virtual-earth arrangement, the voltage of the inverting input remaining close to that of the input (figure 4.6(b)). The voltage gain can be shown to be positive and of value $(R_1 + R_2)/R_1$. The input impedance is increased by the feedback and can thus be greater than the open-loop value, itself often very high. This configuration has the advantages that it does not invert the signal and that it can have very high input impedance. The only disadvantages are that for low gains (such as unity) and large input signals it may exhibit very slight non-linearity and that it is not capable of gains less than unity.

(iii) Differential amplifier

The differential amplifier (figure 4.6(c)) is widely used in transducer applications. It is effectively a combination of the inverting and non-inverting amplifiers, and has a differential gain of $+R_2/R_1$. The input impedance at the non-inverting or positive input is $(R_1 + R_2)$ but the input current at the inverting input is a function of the voltages at both the inverting and non-inverting inputs and is thus more difficult to define. The differential input impedance, however, is simply $2R_1$. The common-mode gain (V_{out}/V_{in} where $V_{in} = V_{in}+ = V_{in}-$), which is ideally zero, depends on the tolerance of the resistors used. It is in the range

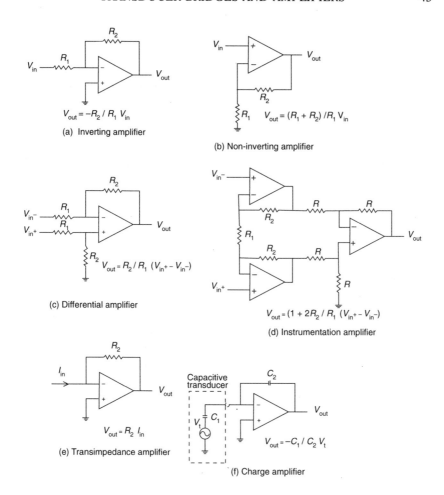

Figure 4.6 *Typical transducer amplifiers.*

$\pm 2p/100$ for $R_2 = R_1$ and $\pm 4p/100$ for $R_2 \gg R_1$, where p is the percentage tolerance of the resistors used. If very close tolerance resistors are used, the common-mode gain of the op amp itself becomes significant.

(iv) Instrumentation amplifier

The instrumentation amplifier (figure 4.6(d)) is perhaps the best configuration to use in many applications. It consists of the differential amplifier described in section (iii) preceded by two non-inverting amplifiers with coupled feedback.

This coupling gives a differential gain of $(1 + 2R_2/R_1)$ while keeping the common-mode gain equal to 1. The complete amplifier thus has a differential gain of $(1 + 2R_2/R_1)$, a common-mode gain of between $\pm 2p/100$ (where p is the percentage tolerance of R) and a very high input impedance. If very high gains are required the differential amplifier can be arranged to contribute some gain as shown in section (iii). Using this configuration, it is easy to construct amplifiers with gains of many thousands and common-mode rejection ratios in excess of 100 dB.

Both differential and instrumentation amplifiers are available complete as integrated circuits. These offer simplified circuitry and greater precision but at a higher cost than the discrete circuits.

(v) Transimpedance amplifier (current-to-voltage converter)

This is similar to the inverting amplifier described in section (i) except that the input resistor R_1 is omitted (figure 4.6(e)). The input current thus flows directly into the virtual-earth point and is balanced by an equal current flowing out through R_2 to the output. The output voltage V_{out} is thus $-I_{in}R_2$ and the input impedance is very low (R_2/A where A is the open-loop gain). This circuit is useful where the output of the transducer is a current, for example a photodiode. It allows an output voltage to be derived while maintaining a constant voltage across the transducer.

(vi) Charge amplifier

This circuit basically gives an output voltage proportional to the integral of current with time, hence charge. It may be used with capacitive transducers and has the advantage that since the feedback is capacitive, the response with respect to the virtual transducer voltage V_t is independent of frequency (figure 4.6(f)). Considered in this way the circuit is inverting and the gain is given by the ratio of the two capacitances and is $-C_1/C_2$. The resistor R_2 is necessary to ensure proper biasing of the op amp at d.c. and adds a low-frequency roll-off below the corner frequency given by R_2 and C_2. This can give problems when the circuit is being used with piezo devices such as accelerometers at low frequencies. In these cases R_2 must be very large which can cause practical problems with the op amp input bias current (see section 4.3).

4.3   Practical operational amplifier characteristics

We mentioned in section 4.2 that an ideal op amp has infinite gain, wide bandwidth, very high input impedance and negligible output impedance. Practical op amps, however, have characteristics that differ significantly enough from these ideals to affect instrumentation circuits. There are a number of characteristics that are particularly significant to the designer of instrumentation circuitry

(i) Open-loop voltage gain (A_{VO})

This is usually quoted in decibels which have the symbol dB ($= 20 \log_{10}\{$voltage gain$\}$). Typical values cover a very wide range (50–150 dB) but this value refers only to the d.c. and very low-frequency gain and tells us nothing about the high-frequency performance. For d.c. excited systems, the open-loop voltage gain should be much greater than the required closed-loop gain. If this is not so then changes in the open-loop gain (for example, due to temperature variations) will affect the closed-loop gain. The greater the ratio of open-loop to closed-loop gains, the more stable the value of the closed-loop gain.

(ii) Gain bandwidth product (GBW)

This figure gives the product of the open-loop voltage gain with frequency. Dividing this figure by the bandwidth required of the closed-loop amplifier gives a guide to the amount of open-loop gain available (see section (i)). Some op amps use second-order compensation techniques which means that the gain bandwidth product is frequency-dependent, being lower at higher frequencies (the low-frequency value is often the only one quoted). Measurement systems that use a high a.c. excitation frequency may need amplifiers with a high gain bandwidth product if they are to produce accurate results.

(iii) Input bias and offset current (I_B and I_{OS})

All bipolar op amps require a small current to flow at the input pins in order to function correctly. Field-effect transistor (FET) input op amps do not require any current but suffer from leakage currents which also flow at the input pins. The average current flowing at the two inputs is called the input bias current, while the difference between the currents at each input is the input offset current. Output offset voltages may be caused by these input currents flowing through external resistances, hence causing input offset voltages. These offsets may be minimised by matching the d.c. resistances at each input of the op amp. For equal input currents this will cause equal offset voltages which will be cancelled by the differential nature of the op amp. Input offset currents will, however, still give rise to output offset voltages which are likely to vary with temperature. If d.c. amplification is required with high source resistances then the op amp used should be chosen for low input offset current.

(iv) Input offset voltage (V_{OS})

A perfect op amp would give an output voltage of zero volts if its inputs were both at the same voltage. Practical op amps, however, require a small voltage across the two inputs for this to be so. This voltage is known as the input offset voltage and can be considered as a voltage generator in series with one of the inputs of the op amp. Offset voltages at the output of the amplifier may be minimised by use of the offset null pins of the op amp (if provided). As with the input bias and offset currents, the input offset voltages are temperature-dependent, and so it is better to use an op amp with a low offset voltage than to try to null one with a higher value.

Note: 8-pin op amps are not necessarily compatible in the function of the offset null pins and damage may occur if different devices are swapped without modifying the circuitry suitably.

(v) Common-mode rejection ratio (CMRR)

Perfect op amps respond only to the difference in voltage between the two input pins, and not to any common-mode voltage present at both inputs. Practical op amps, however, do respond to common-mode voltages and the ratio of differential gain to common-mode gain is known as the common-mode rejection ratio (normally quoted in dB). In the case of the differential amplifier circuit introduced earlier, the limit to the common-mode gain, if the resistors are perfectly matched, is 1/CMRR or –CMRR in dB.

(vi) Slew rate and full power bandwidth (SR)

All practical op amps have a limit to how fast the output voltage can change. This maximum change in output voltage with time is known as the slew rate limit. If the amplifier is to handle high-amplitude, high-frequency signals then it must have a high slew rate. The maximum frequency at which the amplifier can still produce an output from rail-to-rail (usually ±15 V) is sometimes quoted as the full power bandwidth. Although this limit is often much greater than that required by instrumentation type circuitry, some instrumentation amplifiers designed for d.c. have very low slew rates. These types of amplifier can be unsuitable for use in a.c. excited systems, particularly if the excitation frequency is high.

(vii) Stability and uncompensated op amps (A_{VCL})

The majority of op amps available can be used in circuits with any value of closed-loop gain. There are some op amps, however, which have been specifically designed for use as high gain amplifiers. Some of these op amps are not suitable for use in circuits with low values of closed-loop gain and are known as uncompensated op amps. If the closed-loop gain is too low these op amps will become unstable, often giving high-amplitude, high-frequency outputs regardless of their inputs. Care must be taken when using these uncompensated op amps as transimpedance amplifiers since the stability of the circuit will depend on the source impedance, which may not be constant.

(viii) Input noise voltage and current densities (e_n and i_n)

The noise equivalent circuit of an op amp is given in figure 4.7.

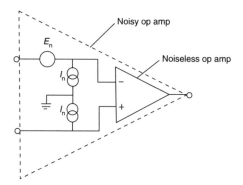

Figure 4.7 *Noise equivalent circuit of operational amplifier.*

The voltage generator E_n and two current generators I_n produce random outputs over a very wide frequency range. Each consists of two components, white noise and $1/f$ (one-over-eff) or flicker noise. The white noise components have spectral densities of e_n V/Hz$^{1/2}$ and i_n A/Hz$^{1/2}$ respectively. The $1/f$ noise components dominate at low frequencies below the $1/f$ corner frequencies f_{ce} for the voltage noise and f_{ci} for the current noise. The values for E_n and I_n over the frequency range f_L to f_H may be calculated as follows:

$$E_n = e_n(f_{ce} \ln(f_H/f_L) + f_H - f_L)^{1/2} \text{ V}_{rms} \qquad (4.1)$$

$$I_n = i_n(f_{ci} \ln(f_H/f_L) + f_H - f_L)^{1/2} \text{ A}_{rms}$$

If the circuit is being used at d.c., the value of f_L might be considered to be zero, hence giving infinite noise. In fact, the time over which the measurement is taken will always give a lower limit to f_L.

If no values for e_n, i_n, f_{ce} and f_{ci} are given by the manufacturer, it may be assumed that the noise performance of the op amp is poor. For reference, the values for the 741 op amp are $e_n = 20$ nV/Hz$^{1/2}$, $i_n = 0.5$ pA/Hz$^{1/2}$, $f_{ce} = 200$ Hz and $f_{ci} = 2$ kHz. FET type op amps do not suffer from $1/f$ noise in their current noise but only in their voltage noise and so no value is given for f_{ci}. In the case of an FET op amp, equation (4.1) may thus be simplified to

$$I_n = i_n(f_H - f_L)^{1/2} \text{ A}_{rms}$$

In circuits with low source impedances the voltage noise of the op amp will dominate, while in circuits with high source impedances the noise current flowing through the source impedance will dominate. FET op amps, with their very low input current noise and high $1/f$ corner frequency for voltage noise, are useful for applications in which the source impedance is high. Bipolar op amps, with their low voltage noise and low $1/f$ corner frequencies, are more suited to applications in which the source impedance is low. For more information on noise and low noise design, see Fish (1993).

4.4 Exercises

4.4.1. Draw an equivalent circuit of the system of figure 4.8. Find the overall responsivity assuming that the thermistor has a resistance of 10 kΩ at the temperature of operation, when its curve of resistance against temperature has a slope of 0.5 kΩ per °C.

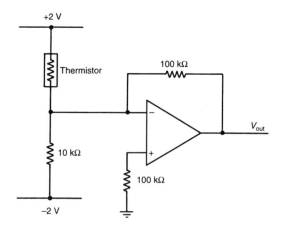

Figure 4.8

4.4.2. A differential capacitive displacement transducer comprises two capacitances in a bridge arrangement. The capacitances are each of 10 pF when in the zero position, and change by ±1 pF per mm. The output may be connected to either a charge amplifier or a non-inverting amplifier, as shown in figure 4.9. Draw an equivalent circuit and find the overall responsivity in each case.

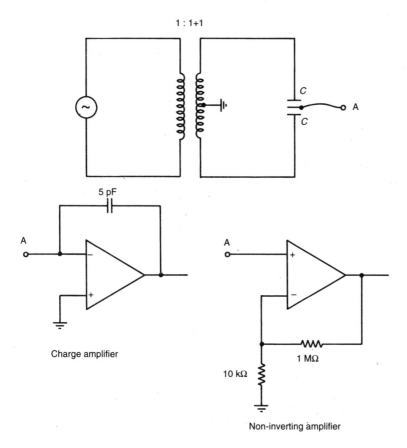

Figure 4.9

4.4.3. Figure 4.10 shows a type 741 inverting amplifier with the input grounded. Find the r.m.s. noise at the output in a bandwidth from 0.1 Hz to 1 kHz, ignoring the noise due to the resistors. Assume that $e_n = 10 \text{ nV/Hz}^{\frac{1}{2}}$, $i_n = 1 \text{ pA/Hz}^{\frac{1}{2}}$, $f_{ce} = 100 \text{ Hz}$ and $f_{ci} = 1 \text{ kHz}$.

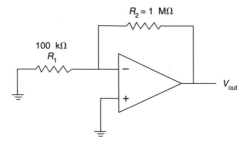

Figure 4.10

5 Transducers for Length

In chapter 3 we discussed the classification of transducers and gave a comprehensive summary of the useful physical effects available. Although such overviews may be valuable and interesting, they are not very helpful in designing or selecting a device for a specific application. The remainder of the book will therefore be devoted to practical aspects. The three most important measurement areas are length, temperature and radiation; in each case we will discuss the basic theory where appropriate, but will concentrate on the design and operation of the most useful devices.

5.1 Classification of length transducers

The standard of length was originally the metre bar, but this was replaced in the 1960s by a definition involving the wavelength of a krypton discharge lamp. This was a big improvement, since in principle any laboratory in the world could set up the apparatus and produce a length standard of high accuracy and reproducibility (about 1 part in 10^8, compared with about 1 in 10^6 for the metre bar). However, it became clear that the most precise measurements possible were those using atomic clocks, and since the velocity of light is a universal constant (299 792 458 m/s) the standard of length was redefined in 1983 as the length of the path traversed by a beam of light in vacuum in 1/299 792 458 second.

We will include under length transducers all length-related quantities, such as displacement, velocity, acceleration and also strain. We will mostly discuss translational devices, mentioning rotational devices only where appropriate. Transducers may be used for relative or absolute measurements. For example, we may wish to detect the relative motion between two objects or we may wish to measure the 'absolute' vibration of an object (for example, in a rocket or spacecraft) where no reference surface is available. However, we will consider only relative transducers in this chapter.

5.2 Displacement transducers

The most important applications for displacement transducers involve ranges from a few micrometres to a few metres. Responsivities vary from about 1 V/m to 10^5 V/m or more, with detectivities as high as 10^{12}/m. The two main classes of displacement transducers are electrical and optical. The former involve one of the primary electrical elements (R, C or L) and the latter mostly involve some form of opaque/transparent pattern on a disc or plate and are often digital.

5.2.1 Electrical displacement transducers

There is an important general formula for the responsivity of a linear electrical displacement transducer. If the impedance of the device varies from zero to Z_{max} over the range l_{max}, and the excitation voltage is V_{ex}, then for a change δZ corresponding to a displacement δl we have an output δV given by

$$\frac{\delta V}{V_{ex}} = \frac{\delta l}{l_{max}} = \frac{\delta Z}{Z_{max}} \quad \text{and} \quad r = \frac{V_{ex}}{l_{max}} \tag{5.1}$$

This can easily be seen to apply to a resistive potentiometer, as shown in figure 5.1.

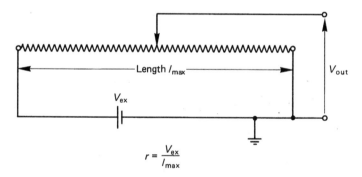

Figure 5.1 *Responsivity of a resistive potentiometer.*

It can be seen from equation (5.1) that one can obtain a high value of r for a given excitation only at the expense of the range, and while this may now be obvious it has often not been realised in the literature.

Resistive displacement transducers

Resistive transducers may be translational or rotational, the simple rotary potentiometer being the most widely used type. They are relatively cheap and may have linearity of 0.1 per cent or better. Their main disadvantages are associated with the necessity of physical contact between the wiper and winding, and some friction and wear are inevitable. In cheap wire-wound devices the gauge of the wire may limit the resolution, but this can be avoided by a suitable design of wiper. A comprehensive treatment is given by Doebelin (1966). Devices are available with ranges up to about 50 cm and rotational types up to 20 or more turns. However, although cheap and useful in simple applications, resistive transducers are unsuitable for most precision applications.

Capacitive displacement transducers

We saw in chapter 3 that the most important capacitive devices are the variable-area and variable-separation transducers. It is not usually convenient to employ only two plates, and differential arrangements with three plates are nearly always used.

(i) Variable-area type   A simple variable-area transducer is shown in figure 5.2.

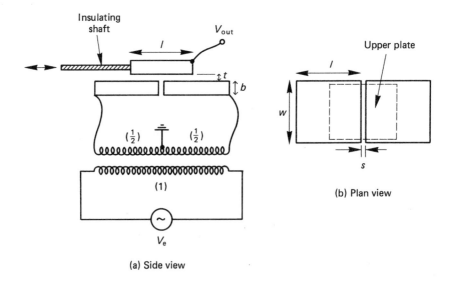

(a) Side view

Figure 5.2   *Variable-area capacitive transducer.*

The device comprises three identical plates, two of them fixed and separated by a small distance s, and the upper one capable of sliding over the fixed plates with a constant separation t. The two fixed plates are excited in antiphase by means of a centre-tapped transformer $(1 : \frac{1}{2} + \frac{1}{2})$. It is clear that when the centre plate is symmetrically placed with respect to the fixed plates no net voltage will be induced on it, whereas movement to the left will produce an increasing voltage of phase the same as that of the left-hand fixed plate, and similarly for movement to the right. The magnitude indicates the total displacement and the phase the sense. The range is clearly equal to the plate length l (that is, $\pm l/2$). The equivalent circuit of the arrangement is shown in figure 5.3.

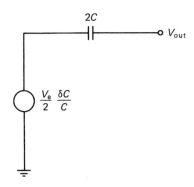

Figure 5.3 *Equivalent circuit of figure 5.2(a).*

The capacitances between the upper plate and the two fixed plates are each C when in the central position. δC is the change in capacitance for a small displacement δx, when the two capacitances then become $C_1 = C + \delta C$ and $C_2 = C - \delta C$. Now

$$C_1 = C + \delta C = \frac{\epsilon \epsilon_0 w}{t} \ (l/2 + \delta x)$$

and

$$C_2 = C - \delta C = \frac{\epsilon \epsilon_0 w}{t} (l/2 - \delta x)$$

giving

$$\frac{\delta C}{C} = \frac{\delta x}{l/2}$$

The responsivity is

$$r = \frac{V_{\text{out}}}{\delta x} = \frac{V_e}{2} \frac{\delta C}{C} \frac{1}{\delta x} = \frac{V_e}{l}$$

equal to the excitation voltage divided by the range, as expected.

It is fairly easy to construct such a transducer. The plates can be 2–3 cm square
pieces (2.54 cm is a good number), cut from 3 mm brass sheet. The distance s is
not critical but should be 1 mm or less, and the separation t should be about
1 mm. This can be arranged by means of an insulated sheet, fixed to the two

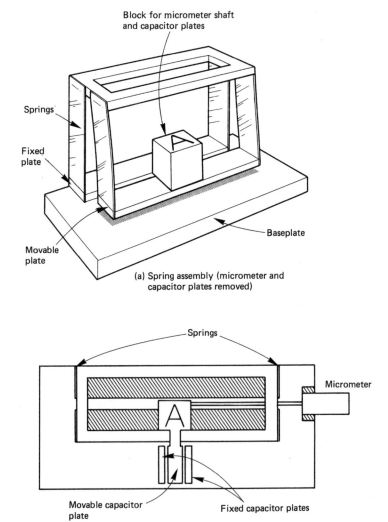

Figure 5.4 *Parallel motion spring mounting (with variable separation transducer).*

lower plates, which can contain guides to ensure correct movement of the upper plate. An alternative method is to use an arrangement of springs, as shown in figure 5.4, which provides smooth parallel motion (though the separation t changes slightly for large deflections).

The transformer is particularly easy to construct. Three 10 m lengths of thin copper wire are twisted together with a hand drill and wound together on a small (1 cm ID) pot core. This ensures that the secondaries are identical but produces a $1 : 1 + 1$ transformer rather than the $1 : \frac{1}{2} + \frac{1}{2}$ described above. For very precise applications it is better to separate the primary from the secondaries. The primary inductance should be about 10 mH and the operating frequency, say, 10 kHz.

A charge amplifier is usually used with capacitive transducers, since the transducer impedance is capacitive; the effective gain, with respect to the voltage generator in figure 5.3, is then independent of frequency. The capacitance C will be about 10 pF with the dimensions quoted above, so the feedback capacitance in figure 5.5(a) must be less than this to produce a reasonable voltage gain. For very precise applications a gain of about 10 is required and this can be obtained from a potential divider arrangement as shown in figure 5.5(b). The resistor R_f is required to provide a suitable bias current and should be such that $R_f \gg 1/2\pi f C_f$ at the operating frequency f of, say, 10 kHz.

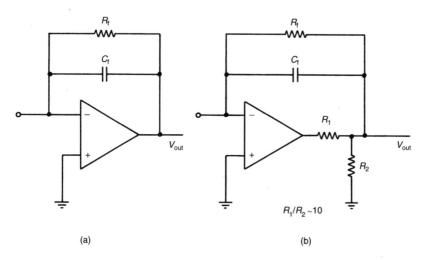

(a)   (b)

Figure 5.5 *Charge amplifiers.*

With excitation voltage 2 V r.m.s., plate length $l = 2$ cm, capacitances $C = 10$ pF and $C_f = 10$ pF, the overall responsivity is

$$r = \frac{2}{2 \times 10^{-2}} \times \frac{20}{10} = 200 \text{ V/m}$$

The output waveforms for displacements of ± 1 mm from the central position are shown in figure 5.6(a), together with the excitation. If these are fed to a fullwave rectifier or DVM the result will always be positive, so an ambiguity will exist regarding the direction of movement. It is necessary to use a phase sensitive detector (PSD), with reference fed from the excitation, to resolve the direction. A PSD is essentially a multiplier so in-phase reference and excitation produce a positive result and antiphase signals a negative result. The output as a function of displacement is shown in figure 5.6(b).

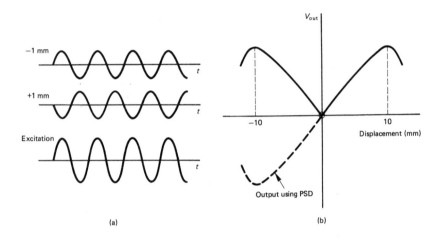

(a) (b)

Figure 5.6 *Output voltages of capacitive transducer.*

The range of the transducer (equal to the plate length) is 2 cm. However, much larger ranges are possible by using a multiplate arrangement, in which several adjacent plates (instead of just two) are each supplied by a different tap on the transformer. Ranges of up to a metre are possible, though with low responsivity since the general formula still applies.

(ii) Variable-separation type A typical variable-separation transducer is shown in figure 5.7.

Again the three plates are identical, but the central plate is constrained to move perpendicular to the fixed plates. The output from the centre plate is zero when in

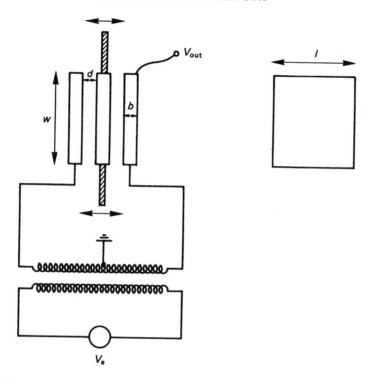

Figure 5.7 *Variable-separation capacitive transducer.*

a central position and increases as it moves left or right. The range is equal to twice the separation d.

The equivalent circuit is the same as for the variable-area type, as shown in figure 5.3. For a displacement d we now have

$$C_1 - C_2 = 2\delta C = \frac{\epsilon\epsilon_0 lw}{d - \delta d} - \frac{\epsilon\epsilon_0 lw}{d + \delta d} = \frac{\epsilon\epsilon_0 lw}{d^2 + \delta d^2} \times 2\delta d$$

and

$$C_1 + C_2 = 2C = \frac{\epsilon\epsilon_0 lw}{d^2 + \delta d^2} \times 2d$$

giving

$$\frac{\delta C}{C} = \frac{\delta d}{d}$$

The responsivity is thus

$$\frac{V_{out}}{\delta d} = \frac{V_e}{2} \times \frac{\delta C}{C} \times \frac{1}{\delta d} = \frac{V_e}{2d}$$

as expected from the general formula. It is interesting to note that the device is inherently linear, although a two-plate version would be very non-linear. However, as shown below, non-linearity is usually observed in practice.

Constructional details of this transducer are similar to those of the variable-area device. The separation d should be about 1 mm to produce a reasonable capacitance of about 10 pF, so the range is very small. Mechanical guides can be made to restrict the movement of the centre plate, but the spring arrangement of figure 5.4 is very convenient.

The transformer and amplifier are similar to those above, and are shown in figure 5.8.

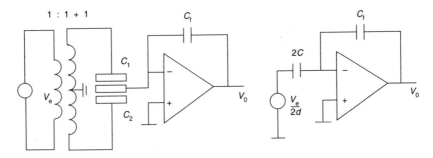

Figure 5.8 *Circuit arrangement and equivalent circuit for variable-separation capacitive transducer.*

With V_e = 2 V r.m.s., d = 1 mm, $2C$ = 20 pF and C_f = 10 pF

$$V_0 = \frac{V_e}{2d} \delta d \times \frac{2C}{C_f} \approx 2000 \text{ V/m}$$

Unfortunately C is not constant, being equal to $\epsilon\epsilon_0 lwd/(d^2 + \delta d^2)$, so the output shows a serious non-linearity. This is because a current (low input impedance) amplifier is used. If a voltage (high input impedance) type was used, one would obtain a linear output. However, the low impedance arrangement has considerable practical applications in insensitivity to stray capacitances and electromagnetic interference, so it is always used in practice. The small range limits the application of the transducer, and it is usually used as a null-sensing device, so that the linearity is of little importance. The form of the output against displacement is as shown in figure 5.9, again using a PSD to obtain a bipolar output, indicating direction of motion. The slope in the central region is that deduced above (2000 V/m).

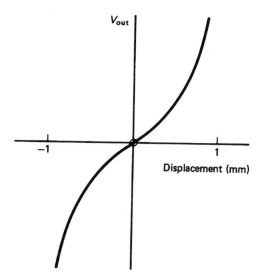

Figure 5.9 *Output voltage of variable-separation transducer (with PSD).*

Inductive displacement transducers

Inductive transducers are widely used in industry, since they are compact and robust. They are less easily affected by environmental factors (humidity, dust, etc.) than capacitive types. Their detectivity is adequate for most applications, though capacitive types are preferable for very precise measurements, since their detectivity is higher and the (unwanted) forces exerted in the measurement are much lower.

(i) Variable-coupling transducers These transducers consist of a former holding a centre-tapped coil, and a ferromagnetic plunger, as shown in figure 5.10.

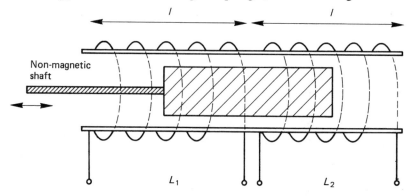

Figure 5.10 *Variable-coupling transducer.*

The plunger and both coils have the same length l. As the plunger is moved the inductances of the coils change; ideally the maximum inductances will be very high and the minimum (when the plunger is in the other coil) quite low. The range is therefore l. The two inductances L_1 and L_2 are placed in a bridge circuit with two equal balancing resistors R, followed by an amplifier and PSD, as shown in figure 5.11.

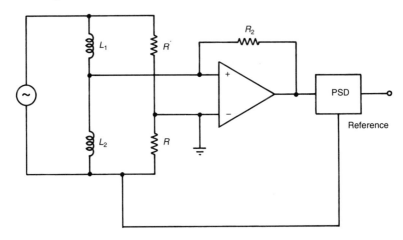

Figure 5.11 *Inductive bridge.*

If the inductances in the central position are each L, corresponding to a plunger length $l/2$, a change in displacement of δl produces changes $+\delta L$ and $-\delta L$, so ideally $\delta L/L = \delta l/(l/2)$, and the output from the bridge is $(V_{ex}/2 \times \delta L/L)$, giving responsivity $V_{ex}/$range as usual. It is particularly easy to construct a transducer of this type. One simply winds a centre-tapped coil on a suitable former, using a rod of mild steel for the plunger. Devices, sometimes known as linear displacement transducers or LDTs, are available commercially up to about 0.5 m in range.

(ii) Differential transformers   These are probably the most widely available displacement transducer, being very robust yet capable of detecting displacements in the nanometre range. They comprise a transformer in which the coupling between primary and a split secondary depends on the position of a ferromagnetic plunger, as shown in figure 5.12(a). The two halves of the secondary winding are connected in series opposition.

There are various designs, but in figure 5.12 the coils all have the same length l, and the plunger has length $2l$. In its central position it protrudes equally into the two secondaries, and in its extreme positions produces maximal coupling with one secondary and minimal with the other so that the range is l. Since the secondaries are in opposition the output is zero in the central position and increases on either side (but with opposite phase).

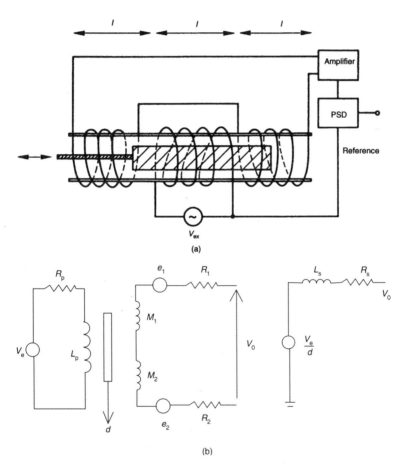

Figure 5.12 *Linear variable differential transformer.*

If we assume that the coupling is perfect in the two extreme positions, then for a unity turns ratio transformer (that is, $1 : \frac{1}{2} + \frac{1}{2}$) the secondary voltage will be equal to half the excitation voltage in these positions, and the responsivity is therefore V_{ex}/l.

An equivalent circuit is shown in figure 5.12(b).The primary has inductance L_p and resistance R_p, and the secondaries have mutual inductances M_1 and M_2 and resistances R_1 and R_2. The equivalent circuit is strictly correct only for perfect coupling and for R_p small, and in practice a frequency-dependent phase shift occurs in the output voltage.

Commercially produced devices have ranges from about 1 mm to 20 cm and are usually known as LVDTs (linear variable differential transformers).

The windings are enclosed in a stainless steel sheath, as illustrated in

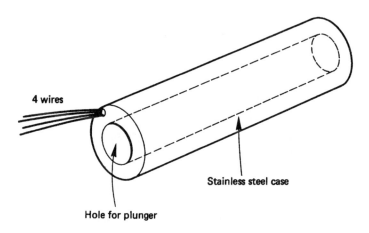

4 wires

Stainless steel case

Hole for plunger

Figure 5.13　*External view of LVDT.*

figure 5.13, with four wires protruding. Two wires are simply connected to an oscillator at, say, 1 kHz and the other two into an amplifier, so the device is very easy to use. The responsivity (unfortunately often called 'sensitivity') is usually given in the form of millivolts per volt excitation per millimetre displacement. The turns ratio is not usually given, so one cannot apply the general formula for responsivity. The output impedance is low, so almost any type of amplifier may be used (virtual earth, non-inverting, differential).

Integrated circuits are now available containing all the excitation and processing electronics needed for an LVDT system, that is, oscillator, amplifier and PSD. This may be supplied built into the LVDT unit, providing a d.c.-energised device. However, for the most precise applications, better results can be obtained with a.c.-energised devices and the associated processing electronics.

It is fairly easy to construct an LVDT, by simply dividing a former into three sections, winding three similar coils, and connecting the secondaries in opposition. Unfortunately such a device will almost certainly have one outstanding characteristic: it will be a very poor LVDT! The problem is that it is not easy to make the two halves of the secondary identical, and their inductance, resistance and capacitance will be different, causing a large unwanted quadrature output in the balance position. Although this can be rejected by a PSD it may seriously limit the usable amplifier gain, and may change with temperature so that attempts to balance it out are hazardous. This is an interesting contrast to the capacitive transducers, where it is easy to make an excellent centre-tapped transformer because the windings may be twisted together. The fact that the two halves are physically separated in the LVDT is an inherent weakness of the device, and precision coil-winding equipment is required to reduce the problem to an acceptable value.

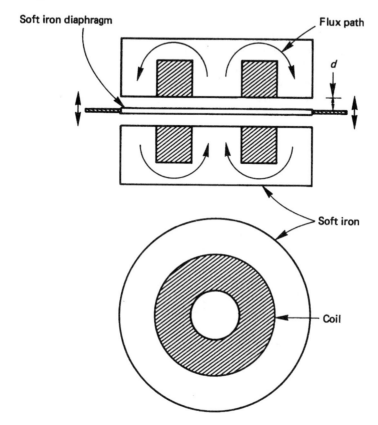

Figure 5.14 *Variable-reluctance transducer.*

(iii) Variable-reluctance transducers These devices can be thought of as the inductive equivalent of the variable-separation capacitive transducer. They have a very small range and are used in rather specialised applications such as pressure transducers. They consist essentially of two closely spaced coils with a thin ferromagnetic disc between them, the position of which changes the inductance of the coils. The disc can be thought of as changing the reluctance of the associated magnetic flux circuit, as illustrated in figure 5.14.

The transducer is operated in a similar way to the variable-coupling type, the responsivity being given by $V_{ex}/2d$, which will be very high since d is usually about 1 mm. Unfortunately the magnetic forces imposed on the disc are quite large, and this limits the application severely. However, the disc can conveniently be a thin diaphragm, hence the application to pressure measurement.

Rotational electrical transducers

Most of the resistive, capacitive and inductive transducers described above are available in rotational form, so the same basic theory applies. There are a few additional rotary devices of interest, notably synchros, which are much beloved of control engineers. They consist of a cylindrical coil which can rotate about an axis, as shown in figure 5.15.

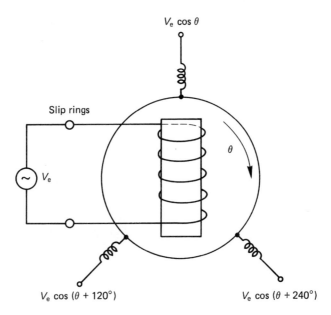

Figure 5.15 *Synchro (schematic).*

A synchro is excited by an a.c. voltage, and surrounded by three coils equally spaced at intervals of 120°. At a given input angle of the central coil, the outer coils detect three voltages that uniquely define the input angle. Synchros have a rather serious disadvantage (this has also been said of control engineers) in that they employ slip rings to feed the central coil, and although convenient in many cases they are unsuitable for high precision measurement.

5.2.2 Optical transducers

Optical transducers have become very popular in recent years, partly because they are mostly inherently digital and partly because they are fairly immune from electrical interference. The main types are encoders, carrying some form of code representing position, and gratings. Encoders may be absolute or incremental, and are usually used for angular measurements, whereas gratings are purely incremental and are used for both translation and rotation.

Absolute angular encoders

A binary-coded absolute encoder is shown in figure 5.16(a). It consists of a number of concentric tracks containing a pattern of opaque and transparent sections such that a unique binary code can be read from the patterns for any angle. The patterns may be read by means of photocells, as shown in figure 5.16(b).

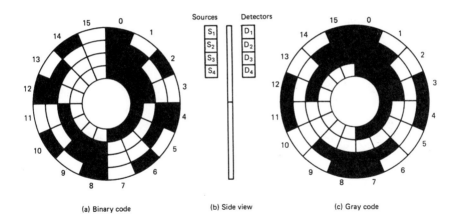

(a) Binary code   (b) Side view   (c) Gray code

Figure 5.16 *Absolute angular encoders.*

One problem with this arrangement, to which some books gleefully devote many pages, is that when several tracks change together (as between positions 3 and 4 in figure 5.16(a)), a serious error could occur if a readout is attempted midway between these positions. The well-known Gray code (figure 5.16(c)) was developed to overcome this, and has the property that only one track changes at a time so whenever the readout occurs it cannot be in error by more than one bit. Simple algorithms exist for converting between Gray code and binary. Absolute encoders are available with 8, 10 or 12 tracks, but become increasingly expensive, and also physically large, as the number of tracks increases.

The need for the Gray code arose because the readout was essentially asynchronous; it could be required at any time and was not related to any reference or clock waveform. However, recent technological developments have made it usual, or at least convenient, to read the outputs of encoders directly via microprocessors, and the earlier difficulties no longer apply. One can either reject an invalid reading occurring at a crossover point (as one would do intuitively) or use an additional clock track to ensure that readings are taken only where valid (the previous value is taken if the disc comes to rest at a crossover on the clock track).

Actually the complication of requiring n tracks for a measurement to an accuracy of n bits is no longer necessary. A rather novel alternative is to use a single track containing a maximal-length binary sequence of, say, 256 digits for 8 bits. Such a sequence contains all the different combinations of 8 digits exactly once each, and every angular readout (determined by 8 adjacent bits) is unique. A separate clock track is usually used, as in figure 5.17, which shows a simple 4-bit version. It is interesting to note that the eight photocells do not have to be adjacent; any equal spacing is satisfactory. In fact, one can actually use only two photocells, reading them serially into the processor, but this requires an initial movement over eight digits to define the first position. A few extra bits can be obtained by interpolation between the clock track digits.

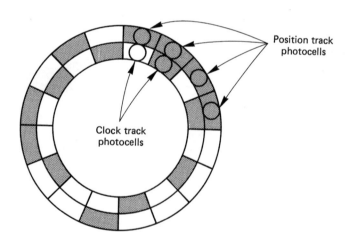

Figure 5.17 *Absolute encoder with maximal-length binary sequence (including the 0000 state).*

Incremental angular encoders

Incremental encoders comprise a disc with a single binary track; this enables them to be made in 'slotted' form rather than transparent/opaque, as for coded discs. The slots are detected by a photocell arrangement feeding a counting system, so the resolution is essentially equal to the angle between the slots. In order to detect direction of rotation, two photocells are placed about a quarter of a slot width apart and the relative phases of the two signals reverse when the direction changes. The signals are fed to a direction-detector logic circuit which controls an up/down counter. Typical discs usually have numbers of slots between about 48 and 96. Figure 5.18 shows a disc and the two photocell signals.

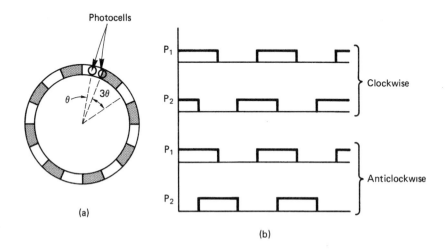

Figure 5.18 *Incremental angular encoder (a) and waveforms produced (b).*

Optical gratings

A grating measurement system can be thought of as a development of a well-known mechanical–optical modulating transducer which we have omitted to mention so far – the ruler. One could observe a grating containing many lines with a microscope and simply count lines, but this would be a little tedious; unfortunately the process is not easy to automate. Grating measurement systems overcome the problem by using two identical equal-mark/space gratings, one fixed (the scale grating) and the other (the index grating) capable of moving over it, as shown in figure 5.19.

As the index grating moves the whole field goes light or dark, so that a large photocell can be used, covering many lines, unlike the situation with the ruler where individual lines have to be observed. This has the added advantage that errors in single lines are automatically averaged out. These systems originally used two gratings with a small angle between them. This produced a geometrical pattern known as a moiré fringe, named after the moiré silk materials used in nightdresses. A study of such patterns is strongly recommended.

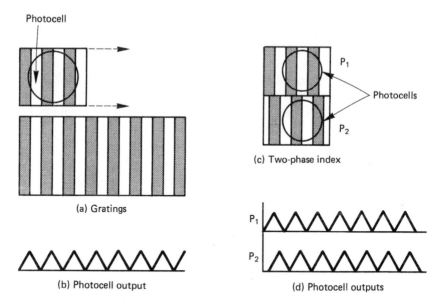

Figure 5.19 *Optical gratings and waveforms produced.*

With a single photocell feeding a counting system there would be no indication of direction of movement. This is usually solved by using a two-phase index, containing two sets of lines displaced by a quarter of a line spacing. Quadrature signals are produced, as in figure 5.18, feeding a direction detector and an up/down counter.

Gratings are available fairly cheaply in a wide range of sizes, for both translation and rotation. The resolution of the measurement is essentially equal to the spacing between the lines, of course, but it is possible to interpolate electronically by a factor of 10 or even 100. This entails essentially finding the ratio of light transmitted in a given position, to the maximum transmitted when coincident, but is actually done by a phase-shifting technique and a set of comparators. This type of electronics is relatively cheap now, so fairly coarse gratings (say 100 lines/cm) are often used in practice followed by suitable interpolation. Difficulties arise in setting up finer gratings, since diffraction of the light falling on the first grating produces a loss in signal, and the gratings have to be placed very close together.

Construction of optical transducers

Most optical transducers are produced photographically, though this may be followed by etching for incremental encoders. The master encoders or gratings are produced by machining, which is of course expensive. However, the availability of microcomputers has made it relatively easy to construct accurate devices, by producing a master on a graph plotter. It is easy to program a PC to draw an A4 size grating via a graph plotter, with an effective resolution of better than 1 mm, and this can then be reduced photographically to the required size. Incremental encoders and radial gratings can be produced very easily, and Gray code (or pseudo-random codes) with a little more difficulty (in programming).

5.3 Velocity transducers

Velocity transducers employ the electrodynamic effect, and are therefore self-generating. There are two main types – moving coil and moving magnet. The former are used mainly in instruments such as accelerometers and velocity meters, whereas the latter are available as off-the-shelf transducers and are very similar to LVDTs.

(i) Moving coil transducers The most usual arrangement comprises a cylindrical magnet in a ferromagnetic yoke with pole-pieces, as shown in figure 5.20.

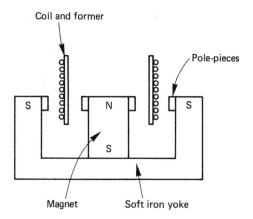

Figure 5.20 *Moving coil transducer.*

The coil is usually wound on a former and attached to the object whose velocity is required. The responsivity is given by $r = Bl$ V/m s^{-1} where B is the induction and l the length of wire cut by the flux. In order to obtain a high value we require both B and l to be large; the air gap has to be small to make B large, so many turns of thin wire are necessary, and in some cases a self-supporting coil (that is, with no former) is used. It is very easy to obtain a responsivity of 1 V/m s^{-1} but in order to obtain a high value (say, 100 V/m s^{-1}) a magnet of about 5 cm diameter and about 10 000 turns of thin wire are necessary. It is not very easy to calculate the responsivity theoretically, but it can be done approximately since for most magnets with air gaps of up to about 5 mm the induction B is about 1 Wb/m^2 and, by calculating the length of wire cut by the flux, one gets a reasonable estimate of the responsivity. For example, with a magnet diameter of 1 cm, a coil of diameter 1.5 cm and pole-pieces such that 100 turns are cut, the effective value of l is

$$2\pi \times \frac{1.5}{2} \times 10^{-2} \times 100$$

so $r \approx 5$ V/m s^{-1} (assuming $B = 1$ Wb/m^2).

As pointed out in chapter 3, the electrodynamic effect is reversible, a current i producing a force Bl newtons, and this can be used to calibrate the transducer. Since the device is usually used with accelerometers in any case, one simply applies a known force to the instrument by adding a mass and balances this by applying a current to the coil, giving $Bl = mg$. The process can of course be done on a beam balance if required.

(ii) Moving magnet transducers The simplest form of moving magnet transducer is just a single coil and magnet, as shown in figure 5.21(a).

This has the disadvantage that the responsivity varies sharply with the exact position of the magnet. In fact, if the magnet was entirely within the coil there would be no net flux cut and no output. A better arrangement is the push–pull device, sometimes known as an LVT (linear velocity transducer), shown in figure 5.21(b), in which the second half of the winding is reversed in sense. This has a much wider range of position over which the responsivity is constant.

The appearance of this push–pull arrangement is very like that of an LVDT. In fact, one can use an LVDT as an LVT by simply applying a d.c. voltage to the primary, which magnetises the ferromagnetic core, taking the output from the (series opposition) secondary.

Rotational velocity transducers

The rotational equivalent of the moving coil velocity transducer is the tachometer. This is another device favoured by control engineers. There are two main types – the d.c. tacho and the a.c. tacho – though both have disadvantages. The d.c. tacho is essentially a d.c. generator with permanent magnet excitation. A coil rotates in the field of a magnet, many poles being used to produce a reasonably smooth output, and slip rings are required to obtain a d.c.

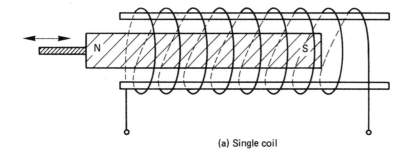

(a) Single coil

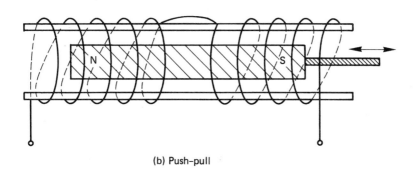

(b) Push–pull

Figure 5.21 *Moving magnet transducers.*

voltage proportional to velocity. The device is very convenient and widely used in control systems where velocity feedback is required, but is unsatisfactory for precision measurement because of unavoidable ripple (due to the finite number of poles) and because of spikes due to the slip rings. The a.c. tacho is better in this respect. It consists of a rotating cylinder and two coils at right angles, one excited at constant frequency and voltage; the second coil detects a voltage proportional to the rate of rotation of the cylinder, owing to eddy current effects. The two transducers are illustrated in figure 5.22(a) and (b).

5.4 Strain transducers

We will discuss piezoelectric transducers and strain gauges in this section. Both these devices are often used for the measurement of force or acceleration, but the actual physical mechanisms involve a change in length so that it is appropriate to discuss them here.

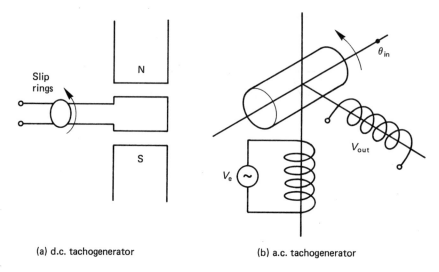

(a) d.c. tachogenerator   (b) a.c. tachogenerator

Figure 5.22   *A.c. and d.c. tachogenerators.*

5.4.1   *Piezoelectric transducers*

Piezoelectric transducers have been with us for a long time and, while such a categorisation is often derogatory, piezo devices have increased in popularity in recent years. As explained above, the basic physical effect is the production of a surface charge density in response to deformation; the magnitude depends on the direction with respect to the crystal axes and is specified by the d and g coefficients (section 3.2.2).

When considered as a length transducer, the physical process is summarised by the equation

$$Q = K\delta t \tag{5.2}$$

where Q is the charge, δt the deformation and K a constant related to the d and g coefficients. Considering a small disc of piezoelectric material of capacitance C, as in figure 5.23, the bulk modulus E of the material is given by

$$E = \frac{F/A}{\delta t/t}$$

so

$$K = \frac{Q}{\delta t} = \frac{dF}{\delta t} = \frac{dEA}{t} = \frac{dEC}{\epsilon\epsilon_0} = \frac{EC}{g}$$

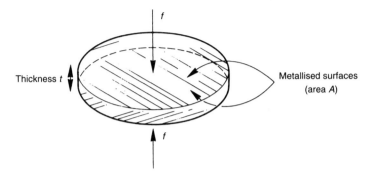

Figure 5.23 *Piezoelectric transducer.*

The equivalent circuit in figure 5.24(a) can be deduced from equation (5.2). It is more useful to choose a current generator i proportional to velocity than a charge generator proportional to displacement; the capacitance C appears in parallel with any leakage resistance which can usually be ignored. The Thévenin equivalent in figure 5.24(b) is also useful, placing C in series with a voltage generator proportional to displacement. For a disc of quartz of area 1 cm^2 and thickness 1 mm, $C \approx 4$ pF, $K \approx 2 \times 10^{-2}$ A/m s^{-1} and $K/C \approx 5 \times 10^9$ V/m, which are very high in comparison with the responsivities found above for displacement transducers, though the applications are not really comparable.

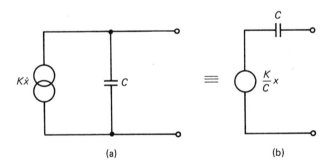

Figure 5.24 *Equivalent circuits of figure 5.23.*

The device can be operated so as to give essentially a response flat to either displacement or velocity, by choice of an appropriate amplifier. For displacement measurements a charge amplifier is used, as shown in figure 5.25(a) with the equivalent circuit of figure 5.24(b). A large resistance R_a is used for bias as with the capacitive displacement transducers. The output V_0 is given by

$$V_0 = \frac{K}{C} x \times \frac{\omega R_a C}{(1 + \omega^2 R_a^2 C_a^2)^{1/2}} \approx \frac{K}{C_a} \times x$$

at high frequencies.

For velocity measurements a current amplifier is appropriate with the equivalent circuit of figure 5.24(a), shown in figure 5.25(b). In this case C_b represents stray capacitance and is only a few pF. The same general expression for V_0 applies, but reduces at low frequency to

$$V_0 \approx KR_b \dot{x}$$

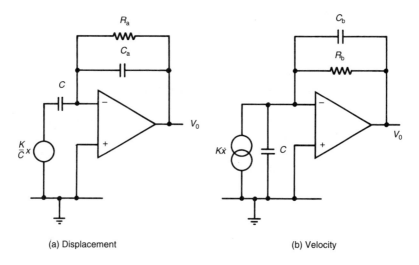

(a) Displacement   (b) Velocity

Figure 5.25 *Amplifier circuits for piezo transducers.*

The two amplifier circuits are of course really the same, but in the first case R_a (for bias) is very high and C_a a few pF, and in the second case R_b is lower and C_b (strays) again a few pF. The two frequency responses are shown in figure 5.26.

A useful velocity response over a wide frequency range (including d.c.) is thus possible and the device finds wide application in miniature velocity meters. However, it is important to note that the displacement response falls to zero at d.c., and although widely used in accelerometers the response usually falls off below a few tens of Hz.

5.4.2 Strain gauges

Strain gauges are essentially thin strips of material whose resistance changes

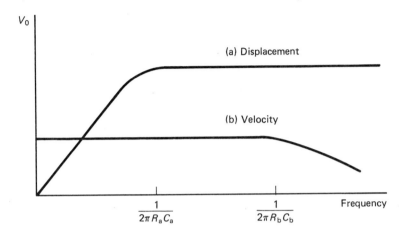

Figure 5.26 *Response of circuits of figure 5.25.*

when strained. We saw in section 3.3.2 that the basic equation of these transducers is given in terms of the gauge factor GF (ratio of fractional change in resistance to strain) as

$$\text{GF} = \frac{\delta R/R}{\delta l/l} = 1 + 2v + \frac{\delta \rho/\rho}{\delta l/l}$$

where v is Poisson's ratio and ρ the resistivity. The second term is known as the dimensional term, and the third as the piezoresistive term.

Strain gauges fall naturally into two types – metallic and semiconductor. For metals v is approximately one-half and the piezoresistive effect is small, so GF ≈ 2, whereas for semiconductors (usually based on silicon) the piezoresistive effect is large (≈ 100) and totally dominant. In fact, $\delta\rho/\rho$ is proportional to stress, so GF is proportional to the bulk modulus E. Unfortunately the high value of GF is accompanied by a high temperature coefficient which limits the application of semiconductor devices.

Metallic gauges may be bonded or unbonded. The latter are simply thin wires (≈ 25 μm) of a material of low temperature coefficient (for example, a copper–nickel alloy known as 'Advance') stretched between pillars, and are sometimes used in cantilever force-measurement systems and accelerometers. Bonded metallic gauges comprise either a wire grid or etched metal foil pattern on a suitable base, which is fixed in position by a suitable adhesive. The choice of adhesive is of some importance because of the requirement of matching temperature coefficients, etc. Various patterns of foil gauge are available, depending on the application required, and some are illustrated in figure 5.27.

The main source of error in measurements with strain gauges is due to

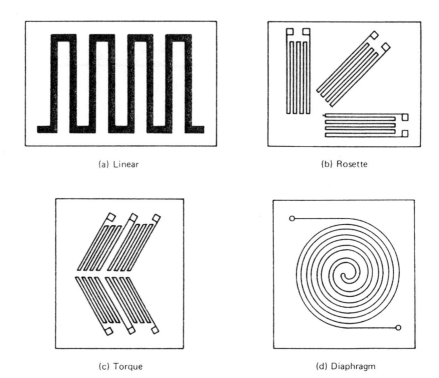

(a) Linear (b) Rosette

(c) Torque (d) Diaphragm

Figure 5.27 *Strain gauge transducers.*

temperature changes, especially with semiconductor devices. Several gauges are used together, some of them unstrained, usually in a four-arm bridge configuration. Two such arrangements are shown in figure 5.28.

A typical amplifier configuration is shown in figure 5.29. The four gauges G_1, G_2, G_3 and G_4 are nominally identical and of unstrained resistance R and connected differentially, so that the equivalent circuit is as shown. The output is directly proportional to strain since $\delta R/R = \text{GF} \times \text{strain}$. It can be seen that

$$V_0 = V_e \frac{\delta R}{R} \times \frac{R_f}{R/2} = V_s \times (\text{strain}) \times \frac{R_f}{R/2} \times \text{GF}$$

The maximum permissible strain is usually about half of one per cent, and it is usual to arrange that the amplifier limits before this value is reached.

5.5 Exercises

5.5.1. Discuss the relative advantages and disadvantages of LVDTs and

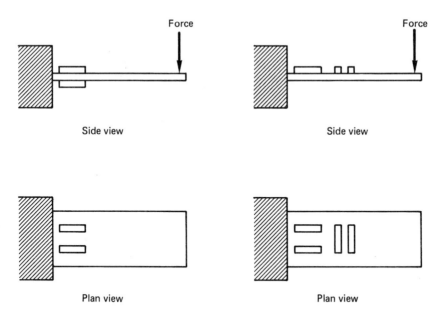

Figure 5.28 *Arrangements of strain gauges.*

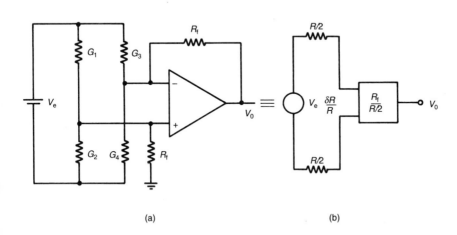

Figure 5.29 *Strain gauge amplifier (a) and equivalent circuit (b).*

variable-area capacitive transducers for the measurement of displacements of less than 1 μm in ranges up to 1 cm.

5.5.2.   Figure 5.30 shows a tiltmeter with a capacitive sensor. The outer plates are excited in antiphase at 1 V r.m.s. at 10 kHz and the centre plate feeds a charge amplifier. The plates have area 10 cm^2 and separation 0.5 mm and the length of the arm L is 5 cm.
(a)   Draw an equivalent circuit of the capacitive bridge.
(b)   Find the (low-frequency) responsivity (in V/radian) at the output.
(c)   Discuss the linearity of the system and how it could be improved.

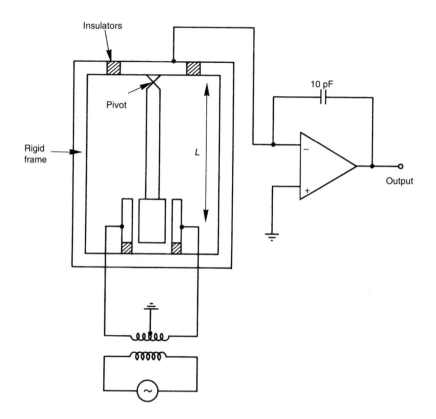

Figure 5.30   *Tiltmeter and capacitive sensor.*

5.5.3.   Figure 5.31 is a schematic diagram of a digital micrometer which is to have a range from 0 to 30 mm and a detectivity of 0.01 mm. Discuss the feasibility of using moiré gratings as the basic sensing element, pointing out possible problems in the mechanical design.

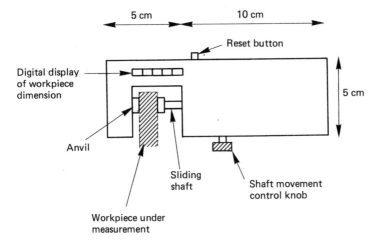

Figure 5.31 *Digital micrometer (schematic).*

5.5.4. Illustrate the terms *responsivity, detectivity, range* and *accuracy* by reference to the applications below, discussing the type of transducer required in each case.

(i) The displacement of the suspended mass in a seismometer is to be measured with an overall responsivity of 10^7 V/m. The maximum displacement is ± 10 μm and changes of $\pm 10^{-12}$ m in a bandwidth of 1 Hz must be detectable.

(ii) It is required to detect displacements of ± 10 μm in a shaft which moves linearly over a distance of ± 1 cm with maximum velocity 1 cm/s. An overall responsivity of 100 V/m is required.

5.5.5. Figure 5.32 shows a schematic strain gauge force transducer, in which force f is applied to the end of a cantilever beam of mild steel (modulus 2×10^{-11} N/m^2). Four metallic strain gauges (gauge factor 2) are attached to the upper and lower faces, as shown, and are connected differentially in a bridge circuit excited at 9 V d.c. and feeding an amplifier of gain 200.·

(i) Draw an electrical equivalent of the bridge circuit and deduce the output of the amplifier for unit strain.

(ii) Find the d.c. responsivity of the system and the maximum measurable input force (assuming that the amplifier limits at ± 4.5 V), and the strain for this force.

Note: A force f at the free end produces surface stress at a distance x from the free end of $6fx/(bd^2)$.

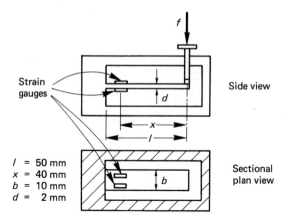

Figure 5.32   *Strain gauge force transducer.*

6 Transducers for Temperature

6.1 Scale of temperature

Although temperature is very apparent to our senses, it is far less easy to define a sensible scale for temperature than for, say, length. This is because we detect or measure temperature by the effect it produces, such as the expansion of a liquid, so an assumption has to be made about the linearity of the effect observed. The temperature of a system is a measure of its energy; this is best described by the Principle of Equipartition, which states that for a system in thermal equilibrium with its surroundings the mean kinetic energy per degree of freedom is $\frac{1}{2}kT$, where k is Boltzmann's constant and T is the absolute temperature.

The most direct realisation of a scale of temperature using this idea is based on the gas equation for an ideal gas, which can be deduced directly from the Principle of Equipartition by means of the Kinetic Theory of Gases. The gas equation $pv = RT$ gives the pressure p of a given mass of gas of volume v at temperature T, with a constant of proportionality R which is equal to the number of molecules multiplied by k. The constant-volume gas thermometer, in which the pressure of a fixed volume of an inert gas is measured as a function of temperature, is the basis of the thermodynamic scale of temperature. This is the nearest one can get to an absolute scale of temperature based on the idea of temperature as a measure of energy. On this scale the only fixed point is the triple point of water, defined as 273.16 degrees Kelvin and as 0.01 degrees Celsius ('by international agreement' is the excuse for the silly number).

Absolute measurements with gas thermometers are made at a few specialised laboratories, but for practical purposes a set of fixed points has been designed, following accurate measurements with gas thermometers, and forms the International Practical Scale of Temperature. The idea is that the various fixed points can be reproduced relatively easily and interpolation between them made by specified transducers. Table 6.1 shows some of the fixed points.

6.2 Temperature transducers

There are not many types of temperature transducer. The two main classes are resistive devices, including both metallic and semiconductor types, and thermoelectric devices. Resistive transducers are modulating, and require a

Table 6.1 Fixed points on the International Practical Scale of Temperature

Temperature (°C)	Description	Example
−182.97	Boiling pt (BP) of oxygen	
0.01	Triple point of water	Platinum resistance thermometer
100.0	BP of water	
444.6	BP of sulphur	
960.8	Freezing pt (FP) of silver	Platinum/Platinum–Rhodium thermocouple
1063.0	FP of gold	

bridge circuit to produce a usable output, whereas thermoelectric transducers are self-generating.

6.2.1 Resistive temperature transducers

These transducers are rather like wire-wound resistors, being in the form of a non-inductively wound coil of a suitable metal wire, usually platinum, copper or nickel. They are often encapsulated in a glass rod in the form of a probe. The normal resistance may vary between a few ohms and a few kilohms, but 100 Ω is a fairly standard value. As seen in chapter 3, the relationship between resistance and temperature is nearly linear, but is more accurately described by equation (6.1):

$$R_T = R_{T_0}(1 + \alpha T + \beta T^2 + \gamma T^3) \qquad (6.1)$$

α is typically about 0.4×10^{-2} for platinum, but β and γ are very small (of the order 10^{-6} and 10^{-12} respectively). The relation is shown in figure 6.3, where it is compared with that for thermistors.

The devices are useful over a temperature range from about −200°C to about 300°C for nickel and copper and up to 900°C for platinum. Their main advantages are that they are stable and reproducible with high linearity, but they are relatively large so have a long time constant. Their responsivity is low, ranging from 0.4 Ω/°C for platinum to about 0.7 Ω/°C for nickel (for a 100 Ω device).

In the last few years improved devices have become available, in which a rectangular matrix of platinum is deposited on a ceramic substrate and connections in the matrix trimmed by laser to produce a resistance of precisely 100 Ω at 300°C. A typical transducer measures 3 cm × 5 mm × 1 mm, and a schematic is shown in figure 6.1.

The transducers are operated with conventional bridge circuits, as shown in figure 6.2, but compensating leads are necessary because of the low element

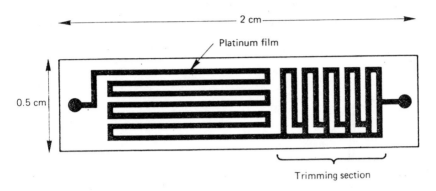

Figure 6.1 *Miniature platinum resistance element.*

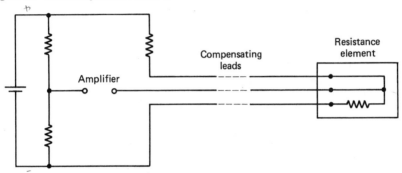

Figure 6.2 *Bridge circuit for resistance thermometer.*

resistance. They cover the range −50 to 500°C and special linearising circuitry is available, providing an output of typically 1 mV/°C.

6.2.2 Thermistors

Thermistors are small semiconducting devices, usually in the shape of beads, discs or rods. They are made by sintering mixtures of oxides of various materials, such as cobalt, nickel and manganese, and are often encapsulated in glass. As we saw in chapter 3, they are inherently non-linear, the resistance following an equation of the form $R_T = Ae^{\beta/T}$ where A and β are constants. The constant A can be eliminated to give the equation

$$R_{T_1} = R_{T_0} \exp\left[\beta\left(\frac{1}{T_1} - \frac{1}{T_0}\right)\right]$$

where R_{T_0} and R_{T_1} are the resistances at temperatures T_0 and T_1. β is usually of value about 3000. The resistance at room temperature may be between a few hundred ohms and several megohms, depending on the type; a typical value is 10 kΩ at 300K with a slope of 1 kΩ/°C. Most thermistors are known as NTC types (negative temperature coefficient) but it is possible to produce PTC types (positive temperature coefficient) over a limited range by suitable doping. Figure 6.3 shows some response curves.

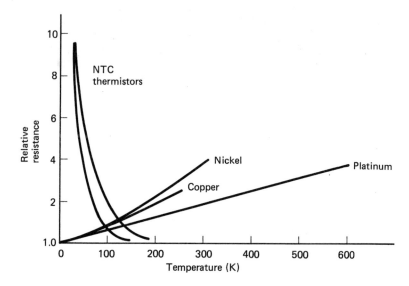

Figure 6.3   *Variation of resistance with temperature for metals and thermistors.*

Thermistors have the advantages of high responsivity and small mass (and hence short time-constant). However, besides being non-linear their characteristics may vary appreciably from sample to sample and they have a limited temperature range of about −100°C to 200°C. Non-linearity is no longer considered much of a disadvantage, since transducers are often connected directly to microprocessors so that the response characteristics can be compensated in software. It is possible to linearise the response by electronic means, which essentially involves placing a suitable resistor in series, and an overall linearity of about 5 per cent over a range between 0°C and 100°C can be obtained (see exercise 6.3.4). However, the most suitable application for thermistors is in temperature control, where their linearity is less important and where their high responsivity enables very high detectivities to be achieved.

Thermistors are usually operated in bridge circuits, and compensating leads are not usually needed because of their reasonably large resistance. Some care has to be taken in limiting the power dissipated, which can seriously affect the measurement because of their small thermal capacitance. Figure 6.4 shows a typical bridge arrangement.

It is easy to analyse the above circuit, by means of the equivalent circuit discussed in chapter 4:

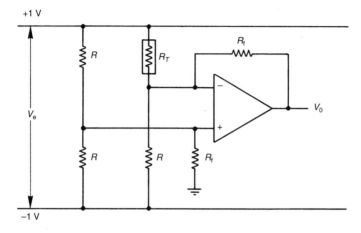

Figure 6.4 *Thermistor bridge circuit.*

$$V_{\text{out}} = \frac{V_{\text{ex}}}{4} \times \frac{\delta R}{R} \times \frac{R_{\text{f}}}{r/2}$$

Worked example

A thermistor has a resistance of 10 kΩ at 300K and the constant β is 3000K. It is operated in the bridge circuit of figure 6.5. Find the responsivity at 300K. Find the output at 310K and deduce the non-linearity with respect to the responsivity at 300K.

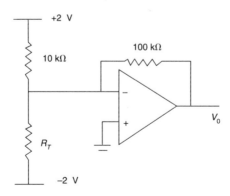

Figure 6.5

$R_T = Ae^{\beta/T}$. At 300K, $e^{\beta/T} = 22026.5$ so $A = 0.4534$.

$\delta R = (-\beta/T^2)Ae^{\beta/T}\delta T$ so $\delta R/R = (-\beta/T^2)\delta T = -1/30$ per $°C$.

Responsivity at 300K $= (V_s/4)(\delta R/R)\left(R_f/\dfrac{R}{2}\right) = 0.667$ V/$°C$.

At 310K, $e^{\beta/T} = 15\ 953$ so $R_T = 7233\ \Omega$

$$V_0 = \left(\frac{2}{10^4} - \frac{2}{7233}\right) \times 10^5 = 7.65\ V$$

Output with 300K responsivity $= 6.67$ V, non-linearity $= 0.98/6.67 = 14.7$ per cent.

6.2.3 Thermoelectric transducers

Thermoelectric transducers are known as thermocouples, and are self-generating devices comprising junctions between two dissimilar wires as described in chapter 3. The conventional arrangement is shown in figure 6.6 where one junction is maintained at 0°C (in melting ice) and the other attached to the object to be measured.

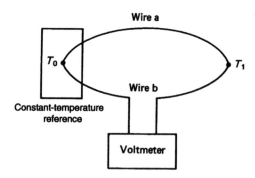

Figure 6.6 *Standard thermocouple arrangement.*

This arrangement is rather inconvenient, because of the layout of the leads and the melting ice container. A more convenient practical arrangement is shown in figure 6.7. The two wires are laid out side by side and are connected directly to the voltmeter. The two junctions between the wires and the terminals of the voltmeter do not produce any error if they are both at the same temperature, but there is no proper reference junction. These two junctions are in fact the reference, so an error tends to occur if the temperature of the room changes. This is avoided by using cold junction compensation, in which the gain of the amplifier is modified so as to cancel the error. A resistance temperature device is included in the amplifier circuitry and changes the gain appropriately.

The arrangement in figure 6.7 is very convenient and almost always used in practice. A flexible stainless steel sheath often encloses the wires to improve the robustness, though at a cost of increased time constant.

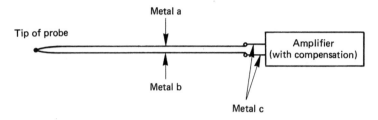

Figure 6.7 *Thermocouple arrangement with cold-junction compensation.*

Thermocouple types are now designated by letters of the alphabet and table 6.2 includes the most widely used devices. The original standard was the Platinum/Platinum–Rhodium device, operating up to high temperatures but with a low output of only about 10 $\mu V/°C$. The Copper/Constantan and Iron/Constantan devices used to be widely used, but the newer base metal types, K and N, are now preferred for many practical applications. One of the disadvantages of thermocouples is their slow response time, but special devices are now available with a response time of only 10 ms (for 63 per cent response to a step change). However, the detectivity of thermocouples is still rather low, owing to their inherently low output, and a detectivity of better than 10/°C is rare.

Table 6.2 Thermocouple types

Type	Name	Materials	Range (°C)	Output ($\mu V/°C$)
J	Iron/Constantan	Fe/Cu–Ni	−25 to 625	60
K	Chromel/Alumel	Ni–Cr/Ni–Al	−200 to 1100	40
N	Nicrosil	Ni–Cr–Si/Ni–Si	−230 to 1230	40
R	Platinum/Platinum–Rhodium	Pt/Pt–Rh	400 to 1500	10
T	Copper/Constantan	Cu/Cu–Ni	−100 to 325	60
W	Tungsten/Tungsten–Rhenium	W/W–Re	up to 2600	20

6.2.4 Solid-state devices

p–n junctions have become popular temperature transducers in recent years. The diode equation relating forward current I to applied voltage V is $I = I_0(\exp eV/kT - 1)$ where I_0 is the leakage current, T the temperature and k is Boltzmann's constant. For low leakage devices this can be written as $V = kT/e \ln(I/I_0)$ so the junction voltage depends on temperature and reduces by about 2.2 mV per °C. Unfortunately, the exact value of the effect varies considerably between samples, so calibration is necessary. However, the effect is very linear and the transducers are extremely cheap.

A typical circuit is shown in figure 6.8. The diodes are usually transistors with base and collector connected together. Strictly they should be operated under constant current but in practice the error is small. Commercial devices are available (such as National Semiconductors LH3911), often with built-in amplifiers.

The same effect is used in miniature encapsulated solid-state devices which are now available. They are sometimes called 'current sources' and are small integrated-circuit two-terminal devices usually calibrated to produce a current of

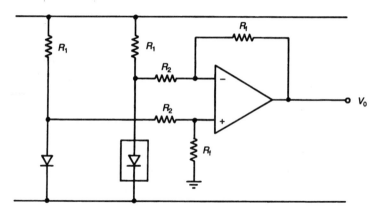

Figure 6.8 p–n *junction temperature transducer.*

1 µA/K, so they produce 300 µA at 300K. They have a working range of about –50°C to 150°C, over which the linearity is about 1 per cent, but unfortunately are rather expensive. Figure 6.9 shows a device arranged to give a responsivity of 0.1 V/°C over the range 0°C to 100°C (by trimming the two resistors R_1 and R_2 respectively). A typical commercial device is the National Semiconductors LM35.

6.3 Exercises

6.3.1. The temperature of the handle of an electric kettle is to be monitored.

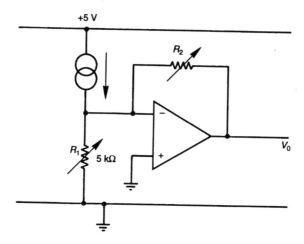

Figure 6.9 *'Current source' temperature transducer.*

Discuss the advantages and disadvantages of the transducers below, recommend the most suitable type and explain how it would be used:
(i)   thermistors
(ii) resistance thermometers
(iii) thermocouples.

6.3.2. The temperatures at six points inside and outside a furnace are to be monitored and recorded on a six-point recorder, sampling each point once every 10 s. The temperature ranges and accuracies required are as follows:
Internal: 3 points in the range 500–1000°C, accuracy ± 5°C
External: 1 point in the range 50–100°C, accuracy ± 0.5°C
        2 points in the range 25–35°C, accuracy ± 0.1°C
Discuss suitable transducers for the system, explaining what instrumentation is required.

6.3.3. A system is required for maintaining the temperature of a small copper cylinder near 50°C, employing a single thermistor as the sensing element. Discuss the sensing part of the system (bridge, circuit, excitation and amplifier) using the details below. The responsivity at the amplifier output should be 1 V/°C, the error due to dissipative heating must be less than 10^{-2}°C, and the detectivity at least 10^2 per °C. [Dissipation constant (power in milliwatts required to raise the temperature 1°C above that of the surroundings): 10 mW/°C. Resistance versus temperature coefficient: 0.4 kΩ/°C at 50°C. Resistance at 50°C: 10 kΩ.]

6.3.4. Figure 6.10 is a circuit for 'linearising' the response of a thermistor, which has a resistance of 10 kΩ at 300K. Find the responsivity at the output at 300K. Calculate the resistance of the thermistor at 310K and deduce the output at this temperature. Calculate the percentage error compared with the output if the responsivity remained constant at the value for 300K. Repeat the calculation at 400K. Assume $\beta = 3000$K.

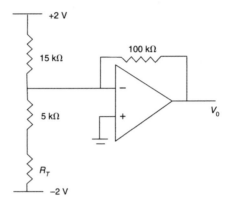

Figure 6.10

7 Transducers for Light

7.1 Light and its properties

Light is an electromagnetic radiation, and visible light lies in a very narrow range of the electromagnetic spectrum. There is no real limit to the latter, it stretches from very low frequency waves of less than 1 Hz, observable by magnetometers, to X-rays and γ-rays of frequencies up to 10^{20} Hz, as indicated in figure 7.1.

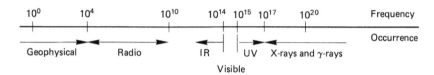

Figure 7.1 *The electromagnetic spectrum.*

Even for visible radiation the frequency is so high (10^{14} to 10^{15} Hz) that no existing photodetectors can respond to the electric or magnetic vector in the electromagnetic wave, and they respond only to the 'time averaged intensity' or mean power of the radiation.

Light can also be considered to consist of packets of energy known as *photons*. The energy per photon is $h\nu$, where h is Planck's constant and ν is frequency. For visible radiation, the energy per photon is very small and most detectors respond only to very large numbers of photons.

The radiation from a hot body, such as a tungsten filament light bulb, is said to be *incoherent*. The excited atoms emit independent short pulses of radiation of differing frequencies, and the total radiant power emitted is simply the sum of that from each atom. Interference effects can be observed only over very short distances (about 1 mm) of the order of the length of the pulse train. The same applies to light from 'fluorescent' discharge tubes and light-emitting diodes (LEDs). In contrast, the radiation from a laser is *coherent*. Lasers employ a process known as *stimulated emission* and successive reflections between the mirrors in the laser cavity cause all the atoms to act together, producing light all of one frequency and with a very large coherence length, so that interference effects can be observed over many metres. Both gas and solid state (laser diodes) are available, but are much more expensive than incoherent sources.

93

Light is often referred to as *radiant flux*, and is measured in watts. However, its effect on the eye varies over the spectral range and is described by the relative luminosity curve, peaking (with value unity) at 0.55 µm. Luminous flux is the brightness evoking capacity of light and the unit is the *lumen*. A radiant flux of 1 watt at 0.55 µm produces a luminous flux of 685 lumens, but at 0.6 µm produces only 411 lumens. Expressed alternatively, 1 lumen at 0.55 µm is equivalent to a radiant flux of 0.00146 W.

Luminous intensity is luminous flux per unit solid angle (steradian) and the unit is the candela (1 lumen/steradian). The *candela* is the formal standard of light, and is defined as one-sixtieth of the luminous intensity of a *full radiator*, which is a black body at a temperature of 2042K (when it looks anything but black), so a luminous intensity of 1 candela at 0.55 µm is equivalent to 0.00146 W/steradian. The brightness of a surface is called *illuminance*, and is luminous flux per unit area. The unit is the *lux*, which is 1 lumen/m^2.

7.2 Classification of photodetectors

The term *photodetector* is usually applied to transducers covering a wavelength range including the visible and near infra-red regions of the spectrum, typically 0.4 µm to about 1 µm. There are many important measurement applications in this range, especially in the near infra-red region, say 0.7 to 1 µm. However for some applications, such as pyrometry (the remote measurement of the temperature of a hot body by collecting the radiation emitted), photodetectors are required for longer wavelengths to 10 µm or more. Pyrometry is discussed in chapter 10.

The responsivity of a photodetector should be expressed in watts, for example 0.5 A/W for a typical solar cell. Manufacturers sometimes give responsivities in lux for visible range photodectors (for example, 7 nA/lux at 560 nm for RS type BPW21), which is actually equivalent to about 0.25 A/W assuming the photodetector area to be about 20 mm^2.

The detectivity of the best photodetectors is similar to that of the eye (about 10^{12}/W). For example, assuming we can just see a 100 W bulb through a 1 mm^2 pinhole at a distance of say 1 km (at night), the radiant flux reaching the eye in the visible region (about one-sixth of the total) is

$$100 \text{ W} \times \frac{1}{6} \times \frac{10^{-6}}{4\pi 10^6} \approx 10^{-12} \text{ W}$$

A flux of, say, 1 µW is therefore very high in a photodetection system.

There are only two main types of photodetector, namely thermal detectors and photon detectors. The former consist essentially of a 'black' screen which ideally absorbs all the radiation incident on it and therefore rises in temperature. The temperature rise is measured by some form of transducer; the photodetectors are usually named after the particular type used, thermocouple and pyroelectric devices being the most popular. Thermal photodetectors usually have a fairly low responsivity but it is essentially constant over a wide range of wavelengths (since all radiation is absorbed, independently of wavelength). This is in sharp contrast to photon photodetectors. These devices employ some form of photoelectric effect, such as the photoemissive effect, the photoconductive effect or the photovoltaic effect, and their responsivity increases linearly with wavelength up to a maximum, beyond which it falls rapidly to zero. This is because the effect occurs only if the photon energy $h\nu$ of the incident radiation is greater than or equal to that of the relevant photo effect, such as work function or energy gap, so the process is inefficient at short wavelengths (since only one photoelectron is produced per photon) and falls to zero beyond a critical wavelength. Table 7.1 summarises the various types.

Table 7.1   Types of photo detector

	Responsivity	*Types*	*Transduction type*
Thermal	Constant	Thermocouple Pyroelectric Bolometer (thermistor)	Modifiers followed by temperature transducer
Photon	Increases with λ up to cut-off	Photoemissive Photoconductive Photovoltaic	Modulator Modulator Self-generator

7.3   Thermal photodetectors

As we saw in chapter 3, a stream of radiation incident on a collecting screen is analogous to the charging of a capacitor by a steady current, and the temperature rise due to an incident stream W watts modulated at frequency f Hz is given by

$$\delta T = \frac{\epsilon W/G}{(1 + 4\pi^2 f^2 \tau^2)^{1/2}} \tag{7.1}$$

where ϵ is the emissivity of the screen, G its thermal conductance and τ the time constant (C/G, where C is the thermal capacitance). We can use this equation to deduce the responsivities of the various thermal detectors.

7.3.1   Thermocouple detectors

Figure 7.2 is a schematic drawing of a thermocouple detector. It is essentially a thermocouple in which the collecting screen is an extended hot junction. The screen is usually made from thin blackened gold foil supported by an insulated ring, and the two contacts are of materials of thermoelectric power of opposite polarity to produce a larger responsivity.

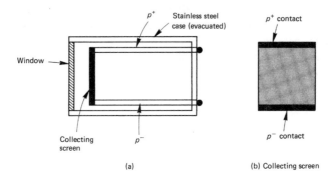

Figure 7.2 *Thermocouple radiation detector.*

The voltage produced for a temperature rise δT is $P\delta T$, where P is the difference in thermoelectric power, so from equation (7.1) the overall responsivity is

$$r = \frac{\epsilon P/G}{(1 + 4\pi^2 f^2 \tau^2)^{1/2}}$$

It can be seen that for large r we require G to be small, and this can be achieved by placing the detector in an evacuated jacket with a window of suitable optical properties. Unfortunately, this makes τ long, so a poor frequency response is obtained. Sometimes the jacket is filled with an inert gas to decrease r, but at the cost of a lower responsivity.

Most thermocouple detectors have a small active area of 1–10 mm^2, a time-constant of at least 10 ms and a fairly low responsivity of 10–100 V/W. The element resistance is typically $100\ \Omega$, making it difficult to obtain a good noise figure with the following amplifier. Their low responsivity and poor frequency response (up to about 10 Hz) limit their application mostly to calibration instruments, such as spectrometers, where absolute measurements of incident radiation are required.

7.3.2 Pyroelectric detectors

These devices form a useful complement with thermocouple detectors, since they are a.c. devices and operate only at frequencies above about 1 Hz. They consist of a thin slice of pyroelectric material, usually lead zirconate titanate, with metallised surfaces and a protective window, as shown in figure 7.3.

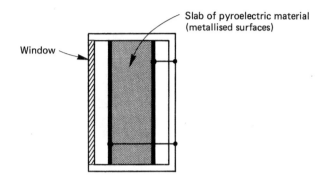

Figure 7.3 *Pyroelectric radiation detector.*

The rise in temperature, δT, is transduced to a surface change density $Q = K\delta T$ where K is a constant. It is more convenient to work in terms of current, which is given by

$$I = \frac{2\pi f \times K \times \epsilon W/G}{(1 + 4\pi^2 f^2 \tau^2)^{1/2}}$$

The responsivity r can be written

$$r = \frac{\epsilon K}{C} \times \frac{2\pi f \tau}{(1 + 4\pi^2 f^2 \tau^2)^{1/2}}$$

This tends to zero at low frequencies but becomes constant and of value $\epsilon K/C$ for $4\pi^2 f^2 r^2 \gg 1$. An equivalent circuit is shown in figure 7.4, comprising a current generator rW in parallel with the electrical capacitance C_1, or alternatively a voltage generator $rW/2\pi f C_1$ in series with C_1.

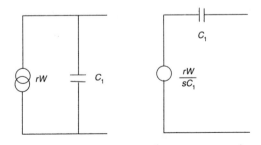

Figure 7.4 *Equivalent circuits of pyroelectric detector.*

Considering the transducer as a current generator it can be connected to a current amplifier as in figure 7.5(a), producing the frequency response shown in 7.5(b).

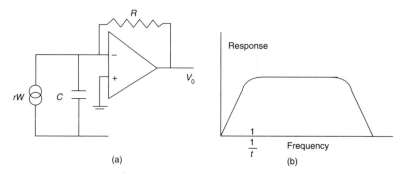

Figure 7.5 *Pyroelectric detector circuit (a) and response (b).*

Typical devices have active areas between 1 and 10 mm^2, thermal time-constants τ about 100 ms, capacitance 10–100 pF and responsivity about 10^{-4} A/W. The frequency response, when operated as in figure 7.5(a), is flat from a few Hz to a few kHz. The windows used limit the spectral response to about 1–10 μm.

Pyroelectric devices are now available with an integral FET amplifier, whose high input impedance significantly affects the response, and it is then best considered as a voltage generator. Such a device and its response are shown in figure 7.6(a) and (b). The response peaks at 1 Hz or less and then falls off as $1/f^2$.

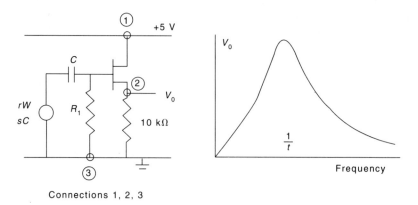

Figure 7.6 *Integrated-circuit pyroelectric detector circuit (a) and response (b).*

For intruder detection, the devices usually have two elements with two attached Fresnel lenses providing different fields of view. This tends to reduce the possibility of a burglar creeping very slowly into a property, making use of the fact that the device does not respond to d.c.

7.3.3   Other thermal detectors

The two devices described above are the most widely used, but a few other types have been developed. The Golay cell was developed many years ago and employs the unlikely sounding principle of expansion of a small flexible cell of inert gas. It is comparable in characteristics to the thermocouple detector. The thermistor bolometer is used in some applications, and uses a thermistor to detect the temperature rise of the collecting screen. A superconducting version has an exceptionally high responsivity and detectivity, though it is used only in special circumstances of course (for example, if the establishment has a lot of money).

7.4   Photon detectors

7.4.1   Photoemissive detectors

Photoemissive detectors are rather like radio valves in appearance, which the writers understand to have comprised an anode and cathode inside an evacuated jacket and to have been used in early radio sets before the advent of transistors and integrated circuits. In figure 7.7 the cathode is composed of a suitable metal which emits electrons when radiation W is incident, provided that the photon energy $h\nu$ is greater than the work function ϕ, and these are drawn off to the anode, producing a photocurrent I given by

$$I = \frac{eqW}{h\nu}$$

where e is the electronic change and q an efficiency factor, known as the quantum efficiency. The current flows through a load resistance R, producing an output voltage which may be amplified as required. The responsivity can be deduced in terms of current or voltage, and is

$$\text{(current)} \quad r_i = \frac{eq}{h\nu} \quad \text{A/W}$$

$$\text{(voltage)} \quad r_v = \frac{eqR}{h\nu} \quad \text{V/W}$$

Practical quantum efficiencies are rather low, often as small as 0.01, but the responsivities obtained are nevertheless fairly high. With a load resistance of 1 MΩ, r_v is about 10^4 V/W at 1 μm. In fact, 1 μm is about the largest usable

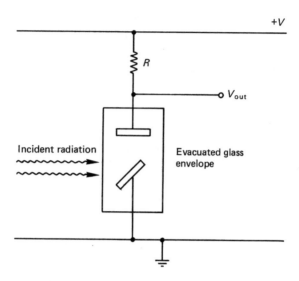

Figure 7.7 *Photoemissive detector.*

wavelength. Very low work functions cannot be obtained, and a combination of caesium oxide and silver is the best-known cathode material.

Photoemissive detectors have the rather serious disadvantage of being rather bulky and fragile, and of requiring fairly high voltages of 100 V or more, though they do have very short time-constants, as low as 10^{-8} s. They are not much used in the form described above, but the photoemissive effect is used in photomultipliers which are still very widely used. In fact, photomultipliers still have a better detectivity (at wavelengths up to 1 μm) than any of the more recent photoconductive or photovoltaic devices. They consist of a photoemissive cathode and a set of additional electrodes known as dynodes at progressively higher voltages. The beam of photoelectrons from the cathode is directed to the dynodes in turn, and extra electrons are produced by secondary emission (from the energy of the colliding electrons), so substantial amplification of the original photocurrent occurs by as much as a factor of 10^6. The particular advantage is that the amplification is obtained prior to the load resistance, in which Johnson noise usually determines the detectivity. By cooling the detector to reduce thermally excited emission a very high detectivity can be obtained. The device is used in very low light levels and in applications where the high excitation voltage required (which may be in the kilovolt region) can be tolerated.

7.4.2 Photoconductive detectors

Photoconductive detectors are composed of semiconductor materials in which incident photons cause excitation of electrons across the energy gap, leading to a change in conductivity. It is important to note that the excited electrons have a finite lifetime so that, when illuminated by a steady beam, an equilibrium condition is rapidly reached in which the decay of the increased number of electrons in the conduction band just matches the excitation. The essential action is thus an increase in conductivity over that in the absence of illumination, and it has been shown previously that the fractional change in bulk resistance is given by

$$\frac{\delta R}{R} = \frac{qW\tau}{h\nu N}$$

where q is the quantum efficiency, W the incident radiant power at spectral frequency ν, τ the lifetime, h Planck's constant and N the number of electrons in the conduction band in the absence of radiation. N is governed by thermal excitation, being given by

$$N = N_0 \exp(-E_g/2kT)$$

where N_0 is the total number of electrons and E_g the energy gap. N therefore rises with temperature and will be relatively large at a given temperature if E_g is small.

Photoconductive detectors are modulating transducers and are operated in a bridge circuit. Figure 7.8 shows a typical arrangement and its equivalent circuit, and it can be seen that the responsivity is given by

$$r = \frac{V_{ex}}{4} \times \frac{q\tau}{h\nu N} \tag{7.2}$$

(The condition where load resistance R = cell resistance R_c gives the optimum responsivity.)

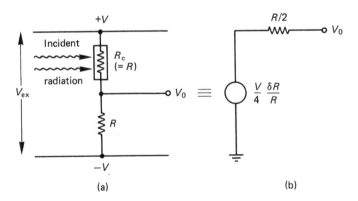

Figure 7.8 *Bridge circuit for photoconductive detector (a) and equivalent circuit (b).*

The responsivity therefore rises linearly with increasing wavelength up to a maximum when $h\nu = E_g$ (that is, $\lambda = hc/E_g$ where c is the velocity of light) and then falls rapidly to zero. Unfortunately equation (7.2) contains several terms whose exact values are not easy to determine, such as q, τ and N, so that although we can deduce the characteristics of these detectors, we cannot find their exact responsivities theoretically. To obtain a high responsivity we clearly require a long lifetime τ, high efficiency q and a small number of electrons N, equivalent to a high resistance R_c in the absence of radiation. If we want a response far into the infra-red the energy gap E_g must be small, so at room temperature N will be high and R_c small. Apparently we cannot have both high responsivity and good long-wavelength response together. Rather surprisingly, since munificence is not often exhibited in physics (and even more rarely in physicists)[*], it is possible to obtain both together at the complication of cooling the detector, usually to liquid nitrogen temperature (77K). This greatly reduces the thermal excitation, increasing the cell resistance considerably and hence the responsivity.

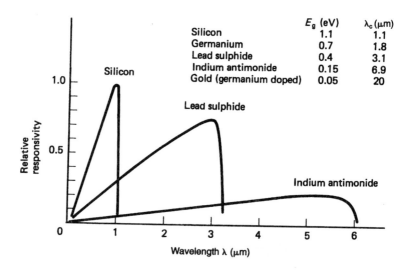

Figure 7.9 *Variation of responsivity with wavelength (room temperature).*

Figure 7.9 shows the variation of responsivity with wavelength for some popular devices, together with their energy gaps and corresponding critical wavelengths. The specific values of responsivity depend on the design of the particular specimen used (area, shape, etc.), since the cell resistance is a controlling factor, so the values shown in the curves are for comparison purposes.

Lead sulphide is one of the best-known general-purpose detectors. It is available in areas of about 1 mm^2 to a few cm^2, usually in the shape of rectangular or circular deposits on a glass of ceramic substrate, as shown in

---

[*]The authors are of course writing in the role of engineers in this chapter.

figure 7.10(a). Cell resistances may be as high as 1 MΩ and time-constants fairly long (≈ 100 μs), so responsivities can be 10^4 to 10^5 V/W but with rather poor frequency response (up to a few kHz). Cooling is not usually necessary.

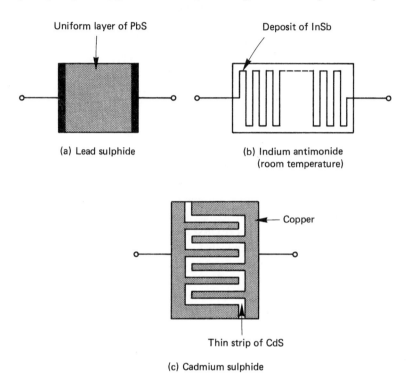

Figure 7.10 *Various photoconductive detectors.*

Indium antimonide is used for more specialised applications. It has the smallest energy gap, and therefore the best infra-red response, of any undoped material, and also has a very short time-constant. It is difficult to use at room temperature, having a resistance of only a few tens of ohms, though such devices are available. However, when cooled it has a responsivity of about 100 V/W and is widely used in remote sensing applications (for example, in thermal scanning of the earth's surface from aircraft and spacecraft) where its good frequency response and excellent spectral response make it ideal. Cooled detectors are similar in appearance to lead sulphide detectors (except that you cannot actually see them because of the attached cooling Dewar). Uncooled detectors are often in the form of a long thin matrix pattern to increase the resistance, as shown in figure 7.10(b).

The effect described above is strictly the *intrinsic* photoconductive effect. *Extrinsic* detectors can be produced by doping, which can place extra energy levels in the otherwise forbidden energy gap, so photons of longer wavelength may produce excitation. Efficiencies tend to be very low and cooling is essential, but it is possible to obtain a response far into the infra-red. Germanium doped with gold responds as far as 20 µm, as shown in the table in figure 7.9.

Another important class of photoconductive detectors employs the charge amplification effect. As explained in chapter 3, impurities can cause hole-trapping effects and result in a very long effective lifetime, though with a similarly increased time-constant of course. The two best-known materials are cadmium sulphide (CdS) and cadmium selenide (CdSe). Both have large energy gaps of about 2 eV, giving critical wavelengths at 0.6–0.7 µm; the response of CdSe matches that of the eye quite well. Detectors using either of these materials are characterised by very high cell resistance (several MΩ) and responsivity (up to 10^7 V/W), and long time-constants (0.1 s). The reproducibility between devices is very poor, since the effect depends on impurities, and they are mostly used in switching applications rather than for accurate measurement. They are often known as light-dependent resistors (LDRs) though their behaviour is very non-linear and the resistance when illuminated may be only a few hundred ohms. In appearance they are at first sight rather like an uncooled InSb device. However, the requirement is exactly opposite; the material has an excessively high dark resistance and in fact the detectors consist of a thin strip of material surrounded by metal on both sides, as shown in figure 7.10(c).

7.4.3 Photovoltaic detectors

The behaviour of a photovoltaic detector is something between a photoemissive and a photoconductive detector. The devices are constructed of similar materials to those used in photoconductive devices, but contain a *p–n* junction which has the effect of causing a physical separation between the hole and the electron pair when subjected to incident radiation, so that a current flows as for photoemissive detectors. As explained in chapter 3, it is the reverse current that is increased when illuminated. In the dark a photovoltaic device is almost indistinguishable from an ordinary *p–n* diode (and its characteristics are the same too), though it has a rather high leakage current. The current/voltage characteristic is the usual diode characteristic, as shown in figure 7.11(a), but when illuminated the whole characteristic moves bodily downwards by an amount equal to the light current I_L. It follows that when operated under short-circuit conditions (for example, when feeding a current-measuring device) the light current I_L is observed, whereas when operated open-circuit (for example, when feeding a voltage-measuring device) a voltage V_{oc} is seen. The transducer is self-generating, since it produces a voltage/current even in the absence of an external power supply or bias, when it is of course the well-known solar cell used for power generation in spacecraft, etc. However, it is usually operated as a modulating transducer at a small reverse bias since the characteristic is more uniform in that region. Also, the capacitance is reduced by reverse bias, improving the frequency response.

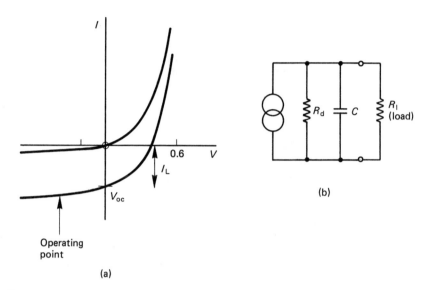

Figure 7.11 *Current/voltage characteristics of photodiode (a) and equivalent circuit (b).*

As shown in chapter 3, the light current $I_L = eqW/h\nu$, so the responsivity is

$$r = \frac{eq}{h\nu} \quad \text{A/W}$$

The efficiencies can be quite high, up to 30 per cent, and the responsivity is typically 0.5 A/W at 1 μm. A useful equivalent circuit is shown in figure 7.11(b), in which the light current is divided between the diode resistance R_d and the load resistor R_l.

The transducer is produced by first depositing a thin layer of, say, *p*-doped material on a suitable substrate followed by a very thin layer of *n*-doped material, forming an extended junction, as shown in figure 7.12. The only common device is silicon-based, normally referred to as the silicon solar cell, and has a critical wavelength of about 1 μm. It is available in a wide range of sizes, from area of 1 mm^2 or less up to several cm^2. The junction capacitance is substantial, often between 100 pF and 1000 pF, and may limit the frequency response.

The transducer may be operated in voltage or current mode, but the latter is usually preferred since the light current is directly proportional to intensity, whereas the relationship between voltage and intensity is not linear. A simple circuit is shown in figure 7.13, though it is capable of excellent responsivity and

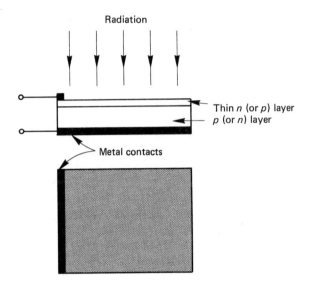

Figure 7.12 *Photodiode detector.*

detectivity. The output to steady illumination can be adjusted by means of the variable resistor. With $W = 2$ µW, responsivity 0.5 A/W at 1 µm and $R_f = 1$ MΩ, the voltage output is 1 V, giving an overall responsivity of 10^5 V/W.

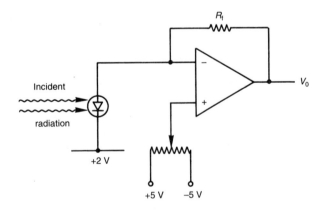

Figure 7.13 *Amplifier circuit for photodiode.*

Photovoltaic detectors are very widely used, being semiconductor devices directly compatible with most electronic circuitry. The detector is sometimes deposited on the same chip as an amplifier, providing a small compact sensor. A large number of devices based on the effect are available.

Phototransistors are similar to ordinary junction transistors but light is allowed to fall on the collector–base region, producing a base current which is amplified by the transistor. These devices are compact and useful in some applications, but they have a poor frequency response and large dark current, and much better characteristics can be obtained by a separate *p–n* junction and amplifier.

Duodiodes are essentially phototransistors without a base connection. They are a *p–n–p* or an *n–p–n* sandwich and act rather like (non-linear) photoconductive devices. They have the interesting property that the polarity does not matter, which is very valuable when students have access to them. They are often of a cylindrical construction of small diameter with a lens in the end, and are used in card readers, bar-code readers, etc.

Pin photodiodes are *p–n* devices with an extra layer of intrinsic (undoped) material between the *p* and *n* layers. This increases the junction area, so increasing the efficiency (and hence responsivity), decreasing the capacitance (and improving the frequency response) and reducing the reverse current (increasing the detectivity). They are very valuable in low light-level applications and where rapid response is necessary.

Photofets are FETs with light allowed to fall on the gate region, and are noted for their large gain–bandwidth product. Avalanche photodiodes employ a breakdown multiplication which produces high gain and low noise, rather like the effect in photomultipliers described above. However, these two devices tend to be expensive and are used only in specialised applications.

Photodiode quadrants are available, comprising four separate devices attached in quadrants for control or sensing applications in which a light beam is directed towards the centre. Linear arrays of 16 or more photodiodes on a 1 mm pitch are used for linear sensing applications. Larger arrays (64 or more) usually consist of charge-coupled devices, rather than photodiodes, in which the accumulated charge is proportional to the total exposure (intensity × time).

7.5 Exercises

7.5.1. An intensity monitor is required for measuring illumination levels in a factory. Discuss the advantages and disadvantages of the following transducers, recommend the most suitable type and explain how it would be used:
(i) cadmium sulphide photoconductive cell
(ii) silicon solar cell
(iii) pyroelectric cell.

7.5.2. Discuss the types of photodetector that could be used for each of the applications below, recommending the most suitable type in each case:
(i) a lamp that switches on automatically when darkness falls (you may assume that it does not switch off automatically as soon as light is present!)

(ii) an intruder-detection system, operating by collecting the radiation emitted by the intruder
(iii) a photographic exposure meter
(iv) measurement of the stability of intensity of a continuous laser beam (mean power about 5 mW at 0.6 µm)
(v) detection of very small amounts of light released in a chemical reaction, in otherwise total darkness.

7.5.3. A light-emitting diode is distant 2 metres from a photodiode, as shown in figure 7.14, and emits 10 µW of radiation at 1 µm, uniformly distributed over a cone of angle 20°. The photodiode has an effective area of 1 cm² and responsivity 0.1 A/W at 1 µm. Find the signal at the output of the amplifier.

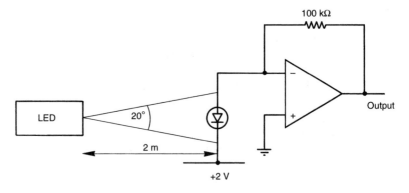

Figure 7.14

7.5.4. A solar radiation monitor is to be designed, in which incident solar radiation falls on a silicon solar cell of area 1 cm² and peak response 1 µm. The overall responsivity is to be 10 V per kW/m². Discuss the design of the system, including a circuit for operating the solar cell. Explain how the intensity/wavelength curve for solar radiation differs from the wavelength response curve for the solar cell, and how errors due to this could be reduced.

(Solar radiation has a maximum value of about 1 kW/m² on a sunny day. The mean responsivity of the solar cell for solar radiation is 0.1 A/W.)

8   Other Transducers

In the previous three chapters we have discussed the basic transducers available for length, temperature and light, the three most important fields. In doing so we have automatically covered nearly all the transducers there are. For example, although we have not specifically discussed force measurement, this is usually done by applying the force to some elastic member and measuring the resulting deflection with a displacement transducer, such as a strain gauge or an LVDT, and we have discussed these latter devices in some detail.

Such measurements usually involve a combination of transducers – a mechanical modifier and a mechanical–electrical modulator in the case of force measurement. Measurements of acceleration, force, pressure and flow all fall into this class, and we will discuss them in this chapter.

8.1   Acceleration transducers

We discussed transducers for relative displacement and velocity under the heading of length transducers, but we totally omitted acceleration transducers. There is a good reason for this – there aren't any! A *relative* transducer makes a measurement with respect to some reference; for example, a relative displacement transducer, such as an LVDT, has one end fixed to an unmoving base. It happens that there are no simple transducers that respond directly to relative acceleration, but surprisingly absolute acceleration can be measured quite conveniently. An *absolute* transducer is one that does not require a reference; for example, if we want to know the accelerations on a driver in a motocross car, we have to use a device that does not require a reference, since clearly no reference (a fixed level surface) is available.

Acceleration transducers use what is known as the principle of seismic mass. They comprise a mass suspended by a spring from a rigid frame; the relative motion of the mass with respect to the rigid frame can be measured with a displacement or velocity transducer and is related to the acceleration of the frame. Figure 8.1 shows a simple accelerometer and its electrical analogue, from which one can easily deduce the transfer function.

The absolute velocity of the rigid frame is $\dot{x}_m$ and the mass has relative velocity $\dot{x}_r$ and absolute velocity (which we do not require here) of $\dot{x}_{abs}$. In the

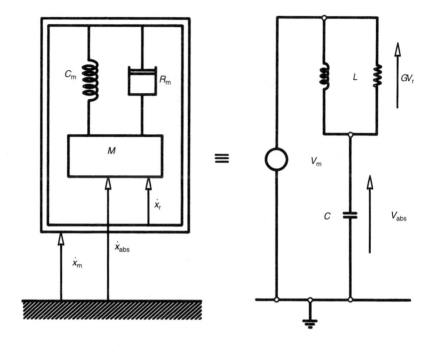

Figure 8.1 *Schematic accelerometer and its electrical analogue.*

analogue circuit the absolute velocity of the mass is represented by the voltage V_{abs} across the capacitor and the relative velocity of the mass is analogous to the voltage V_r across the parallel combination of inductance and conductance.

In terms of the Laplace variable s:

$$\frac{V_r}{V_m} = \frac{\frac{1}{sL}+G}{\frac{1}{sL}+G + \frac{1}{sC}} = \frac{s^2LC}{s^2LC + 1 + sLG}$$

Converting back to mechanical quantities and setting $\omega_0^2 = 1/mC_m$ and $\zeta = 1/2\,mR_m\omega_0$:

$$\frac{\dot{x}_r}{\dot{x}_m}(s) = \frac{s^2}{s^2 + 2\zeta\omega_0 s + \omega_0^2} \tag{8.1}$$

where ω_0 is the natural angular resonant frequency and ζ the damping ratio. Equation (8.1) can be rewritten in terms of relative displacement x_r and input acceleration $\ddot{x}_m$ and becomes

$$\frac{x_r}{\ddot{x}_m}(s) = \frac{1}{s^2 + 2\zeta\omega_0 s + \omega_0^2} \tag{8.2}$$

Equations (8.1) and (8.2) tell us all the important principles of the design of instruments for measuring absolute acceleration, and absolute velocity and displacement as well. The frequency response of the system of figure 8.1 is obtained by putting $s = j\omega$, and is shown in figure 8.2.

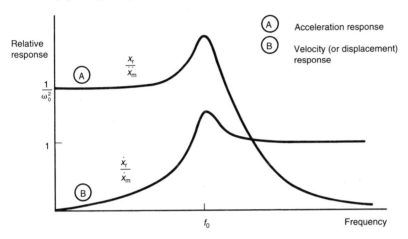

Figure 8.2 *Responses of accelerometer.*

If we measure the relative displacement of the mass, we get an accelerometer that has a flat response (of value $1/\omega_0^2$ up to the resonant frequency (curve A)). If we measure the relative velocity of the mass, we get a velocity meter that has a flat response (of unity) above the resonant frequency (curve B). Alternatively since $\dot{x}_r/\dot{x}_m = x_r/x_m$, curve B shows we can get a measurement of absolute displacement above the resonant frequency by measuring the relative displacement of the mass. The main types of instruments are summarised below.

8.1.1 Accelerometers

All devices are of the same form in principle, the relative displacement of a mass/spring system being measured by a suitable displacement transducer. The design depends very much on the frequency range required, since the response is determined by the natural frequency.

One of the most popular devices is the strain gauge accelerometer shown in figure 8.3(b). It consists of a cantilever beam arrangement with strain gauges as the displacement transducers. Conventional devices have natural frequencies in the range 100 Hz to 1 kHz with ranges up to $\pm 5g$ and can be used in any orientation. Miniature solid-state devices, machined from silicon with semiconductor gauges diffused on to them, are now available, but have higher natural frequencies and correspondingly lower responsivity (proportional to $1/\omega_0^2$).

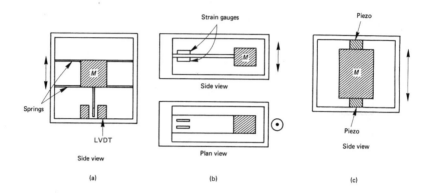

Figure 8.3 *Schematic accelerometers: (a) LVDT, (b) strain gauge, (c) piezoelectric.*

For the highest frequencies, up to 100 kHz or more, piezo accelerometers are very convenient. The mass is supported by one or more piezos, which act both as the spring and displacement transducer as in figure 8.3(c). The devices can be very small but have the disadvantage that their responsivities are very low and they do not respond below frequencies of a few tens of Hz, since piezo devices do not respond to displacement at d.c.

At low frequencies (below 100 Hz) an LVDT or capacitive transducer is used, as shown in figure 8.3(a). The accelerometers have high responsivity and detectivity, and precision devices are used in inertial navigation systems. Such instruments are always operated as feedback systems, in which any tendency to motion of the suspended mass with respect to its supporting frame is sensed by the displacement transducer and a force applied via a magnet–coil actuator to maintain the mass stationary. A schematic system is shown in figure 8.4 and the corresponding block diagram in figure 8.5. The current in the coil is directly proportional to the input acceleration over a range of about ±1*g*.

8.1.2 Velocity meters and displacement meters

There is only one common type of velocity meter, that shown in figure 8.6. The device employs a moving coil/fixed magnet transducer for velocity sensing, and typically has a natural frequency of about 5 Hz (as low as can easily be obtained), the response being flat above this frequency.

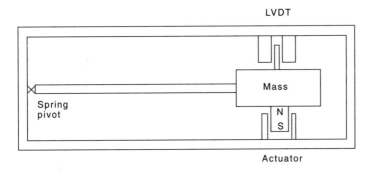

Figure 8.4 *Schematic feedback accelerometer.*

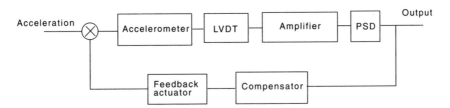

Figure 8.5 *Block diagram of feedback accelerometer.*

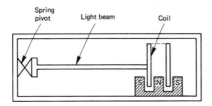

Figure 8.6 *Velocity meter.*

Velocity is most often measured by integrating the output of an accelerometer, when a response is obtained down to zero frequency, but with some integrator drift problems. Displacement can be measured at high frequencies (above resonance) by an accelerometer, but at low frequencies is obtained by double integration of the acceleration. Although this has very serious drift problems, it is the only way that absolute displacement can be measured at low frequencies and is the method used in inertial navigation systems, using accelerometers of the type discussed above.

8.2 Force transducers

The three main methods of force measurement are the mass balance, in which the force is balanced against a known mass, the force balance, in which the balancing force is via a spring or magnet–coil arrangement, and the deflection type, in which the deflection of an elastic element is measured. Mass and force balance systems are often in the form of a beam balance, the deflection being detected by a displacement transducer. The balance is often operated as a feedback system, the tendency for displacement being detected and converted into a current fed to a magnet–coil, producing a balancing force, as shown in figure 8.7. Feedback systems have considerable advantages in linearity and in calibration, since the overall response is governed by the passive feedback element. They are sometimes referred to as *inverse* transducers (see Jones (1977)).

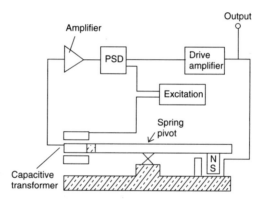

Figure 8.7 *Force balance system.*

8.2.1 *Elastic force transducers*

The elastic element may be in the form of a cantilever beam, pillar, proof ring or diaphragm, and the deflection may be detected by an LVDT, strain gauge or piezoelectric transducer, depending on the range of force or frequency required. LVDTs are used with a cantilever beam for small forces, and strain gauges with a variety of beams and pillars for higher forces. Piezoelectric methods are used for high forces and high frequencies, but again have the disadvantage of not responding at d.c.

A general transfer function for the deflection x of an elastic force transducer in response to a force F is

$$\frac{x}{F}(s) = \frac{\omega_0^2 C_m}{(s^2 + 2\zeta\omega_0 s + \omega_0^2)}$$

where C_m is the compliance, ω_0 the natural frequency and ζ the damping ratio. The frequency response is flat at low frequencies (of value C_m) and falls off as $1/f^2$ above resonance.

Some typical force transducers are shown in figure 8.8.

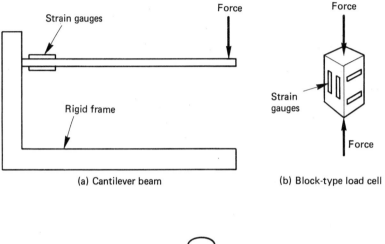

(a) Cantilever beam   (b) Block-type load cell

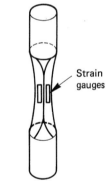

(c) Column-type load cell

Figure 8.8 *Various force transducers.*

8.3 Pressure transducers

The most common methods of pressure measurement are by manometers and mechanical deflection devices. Manometers comprise U-tubes with one end closed, usually containing mercury, the difference in height in the two sides being proportional to the pressure applied at the open end. Deflection transducers may be tubes, diaphragms or bellows, the deflection being measured by a suitable displacement transducer.

8.3.1 *Elastic pressure transducers*

Bourdon tubes have been widely used for pressure measurement. They consist of a flattened tube of approximately elliptical cross-section bent into some shape, such as a 'C' or a spiral, so that the end deflects under pressure. Some examples are shown in figure 8.9. They are fairly linear for small deflections. Bellows are also used and have an improved range and linearity; they are reversible and are useful as displacement–pressure transducers in pneumatic systems.

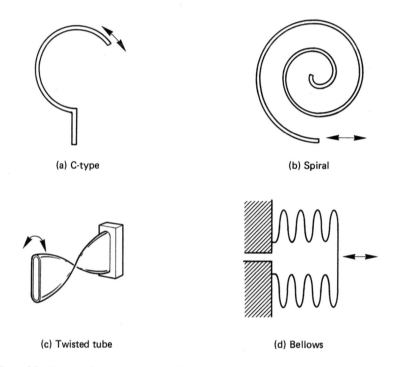

(a) C-type   (b) Spiral

(c) Twisted tube   (d) Bellows

Figure 8.9 *Various elastic pressure transducers.*

Diaphragms are probably the most popular transducer. They consist of a thin stretched membrane whose deflection is usually measured by strain gauges. The deflection of the centre of the membrane, x, is linearly dependent on the pressure applied, p, for deflections up to about half of its thickness, and is given by

$$x = \frac{3(1 - v^2)d^4p}{256Et^3}$$

where v is Poisson's ratio, d is the diameter, E is Young's modulus and t is the thickness.

Transducers for different ranges can be designed by suitable choice of d and t, to keep the deflection for maximum pressure less than $t/2$. For example, with $d = 2$ cm, $E = 10^{11}$ Pa, $t = 0.1$ mm, the deflection $x = 1.4 \times 10^{-8}p$, so a pressure of about 35 bar (1 bar $= 10^5$ Pa) produces the maximum linear deflection of 0.05 mm. Figure 8.10 shows a differential transducer with a soft-iron diaphragm and variable-reluctance displacement transducer.

Modern devices usually comprise a small silicon diaphragm with a semiconductor strain gauge bridge diffused into it, providing a small and compact unit shown in figure 8.11. They may be differential or single ended with ranges up to about 100 bar.

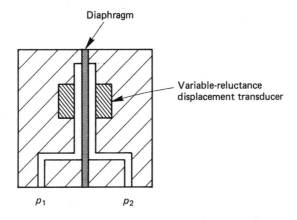

Figure 8.10   *Differential diaphragm-type pressure transducer.*

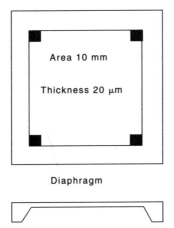

Figure 8.11   *Miniature silicon-diaphragm pressure transducer.*

8.4   Flow transducers

Flow measurement is a large and important subject. Unfortunately it is a rather messy topic, with a large number of different transducers but with little overall unity. It may include vector flow, in which both the magnitude and direction of the flow at a point in space are required, volume flow, which usually refers to the flow in tubes, or mass flow, where the total mass per second is required. We will summarise the methods available for vector flow and volume flow only, since most mass flowmeters involve a volume flowmeter followed by a density measurement.

The flow of the liquid through a pipe may be *laminar* or *turbulent*. In laminar flow each layer slides smoothly over the adjacent layer, producing a velocity profile across the pipe as in figure 8.12(a). In turbulent flow, which occurs at higher velocities, this smooth profile breaks up and the motion and profile is as shown in figure 8.12(b).

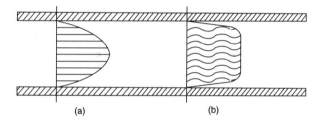

(a)   (b)

Figure 8.12   *Velocity profiles for (a) laminar and (b) turbulent flow.*

The transition between the two is described by the *Reynolds number* R_e, given by $R_e = vd\rho/\eta$, where v is the velocity of the fluid of density ρ and viscosity η, and d is the diameter of the pipe. For $R_e < 2000$ the flow is normally laminar, and usually becomes turbulent for $R_e > 10\ 000$.

One of the most common ways of measuring flow is to change the diameter of the pipe and measure the corresponding change in pressure, and figure 8.13 shows the quantities involved.

For turbulent flow the quantities are related by the *Bernouilli* equation

$$\frac{v_1^2}{2} + gh_1 + \frac{P_1}{\rho_1} = \frac{v_2^2}{2} + gh_2 + \frac{P_2}{\rho_2}$$

which is essentially a statement of conservation of energy.

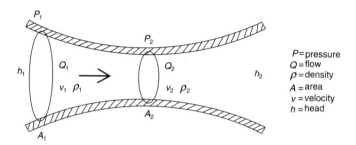

Figure 8.13 *Flow parameters in a pipe.*

For a constant fluid head h and an incompressible fluid: $\rho_1 = \rho_2 = \rho$ and $Q_1 = Q_2 = Q$, so the equation reduces to $\frac{1}{2}(v_1^2 - v_2^2) = (\rho_1 - \rho_2)/\rho$ and the flow rate Q is proportional to $(\rho_1 - \rho_2)^{1/2}$. The coefficient involved is somewhat complex, involving a correction factor known as the discharge coefficient dependent on the type of construction.

8.4.1 Volume flow transducers

In measuring the flow of liquid in a pipe, an ideal transducer would be one that causes no disturbance to the flow, known as a 'non-invasive' transducer and preferably situated externally to the pipe. Unfortunately, very few devices fulfil either of these requirements, though some of the more recent methods do so under some conditions. It has been said that flow experts 'do it non-invasively'.

(a) Plates and nozzles

Most flow transducers are simply modifiers and comprise plates or nozzles that partly restrict the flow and produce a pressure drop. This is exactly analogous to the measurement of electric current by finding the voltage drop across a small series resistor. Some typical devices are shown in figure 8.14. In each case the volume per second is proportional to the square root of the pressure difference, which is measured by a suitable transducer such as that shown in figure 8.10 above.

These transducers are far from ideal, requiring a special insert in the pipe, but are nevertheless very widely used.

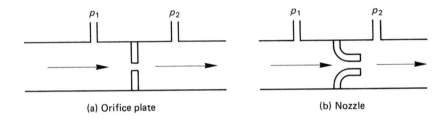

(a) Orifice plate   (b) Nozzle

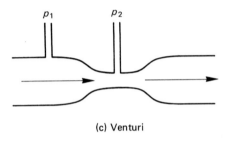

(c) Venturi

Figure 8.14 *Various pressure-drop flow transducers.*

(b) Turbines and positive displacement devices

A turbine is a vaned device placed directly in the flow, and illustrated in figure 8.15.

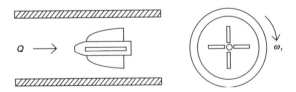

Figure 8.15 *Turbine flowmeter.*

The angular velocity ω_r of the blades is directly related to the flow rate Q, depending on the geometry of the vanes and the area of the pipe. It has the advantages that the disturbance to the flow is small and it may work bidirectionally. In addition, the rate of rotation can be sensed externally to the tube and may be read directly in digital form.

A positive-displacement device is similar, but rotates against a spring rather than rotating freely. A drag-force device employs somewhat similar principles, but is a beam that deflects in the flow.

(c) Electromagnetic flowmeter

The electromagnetic flowmeter is very satisfactory for conducting fluids, and comprises two electrodes in a non-conducting section of tube, as shown in figure 8.16. A strong magnetic field is applied perpendicularly to the plane of the diagram, and the ions are deflected according to their sign and the fluid velocity, and detected by the electrodes placed in a suitable bridge circuit. Apart from the inserted section of the tube, this is almost an ideal transducer.

Various problems, such as polarisation and chemical effects, occur if the field coils are operated d.c. An a.c. excitation at 50 Hz can be used, but better results are obtained by applying low-frequency positive and negative current pulses.

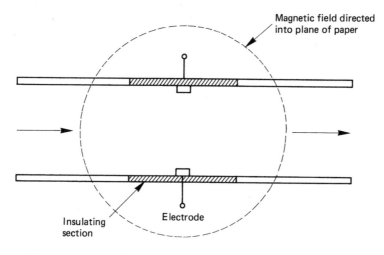

Figure 8.16 *Electromagnetic flowmeter.*

(d) Transit-time methods

Flow velocity can be measured by finding the time taken for an ultrasonic pulse to pass through the liquid to a receiver further downstream, as illustrated in figure 8.17. In the differential transit-time flowmeter, the receiver (R) also emits a pulse after receiving one from the transmitter (T), and the differential time δt is given by $\delta t = 2 D \cot \theta \, v/c^2$, where c is the velocity of sound in the fluid. δt is very small in practice, so the accuracy is limited.

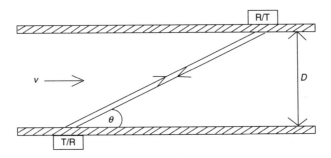

Figure 8.17 *Transit-time flowmeter.*

(e) Doppler methods

Doppler methods provide almost an ideal means of flow measurement, and both optical and ultrasonic methods are available. They involve injecting a beam of coherent light or an ultrasonic wave into the fluid at an angle to the flow, and measuring the frequency shift of the wave reflected from the moving fluid.

A schematic laser doppler flowmeter is shown in figure 8.18. A powerful main beam and a weaker reference beam are passed through the fluid at different angles, and some of the main beam is reflected in the direction of the reference beam by any small eddies or particles in the fluid. These act as doppler reflectors, and the frequency shift δf is given by $\delta f = (2f/c) \cos \theta \, v$, where f is the optical frequency, c the velocity of light, θ the angle relative to the flow and v the flow velocity. Because optical frequencies are so high, the frequency shift can be counted electronically and flows measured over a wide range (a few mm/s to several hundreds m/s).

Optical methods can be used only with transparent fluids, but the equivalent ultrasonic method is not so restricted in its application, though it produces a correspondingly lower frequency shift. Such methods are particularly suitable as 'strap on' flowmeters in industrial applications.

(f) Correlation methods

An interesting recent method, which has ideal characteristics and wide application, is flow measurement by correlation. The principle is that a beam of light or sound is reflected from two adjacent points in the flow (alternatively it may be transmitted) and the correlation function for the two signals $s_1(t)$ and $s_2(t)$ is determined. Provided that the two reflection points are reasonably close, $s_2(t)$ will

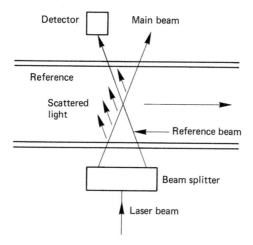

Figure 8.18 *Laser doppler flowmeter.*

be simply a delayed version of $s_1(t)$ and the correlation function will have a peak for a time interval equal to the transit time between the reflection points, which is of course proportional to the velocity of the fluid. Figure 8.19 shows an optical method applied to fluid in an open channel, where the surface ripples produce $s_1(t)$ and $s_2(t)$; in some cases a disturbance (for example, bubbles) must be added to provide something to correlate.

The correlation function $R(\tau)$ is given by

$$R(\tau) = \int_{-\infty}^{+\infty} s_1(t)s_2(t + \tau) \, dt$$

where τ is a time shift. This is not very convenient for calculation but a considerable simplification is obtained by digitising $s_1(t)$ and $s_2(t)$ into binary values. The correlation, though with some reduction in accuracy, can be found rapidly by an on-line microprocessor, since it involves only EOR (exclusive-OR) operations. Indeed, the technique became feasible only with the development of cheap processors, and both optical and ultrasonic methods are widely used. It can be applied in many fields such as measuring the speeds of belts in a machine, the speed of cars or trains, and even the velocity of effluents from remote factory chimneys. Beck (1983) has given a useful review.

(g) Vortex methods

When a fluid flows over a streamlined body the slower moving surface layers remain attached to the body over most of its length. However, if the body is not streamlined (usually then known as a *bluff* body) the surface layers become detached and produce vortices downstream of the body. The vortices are

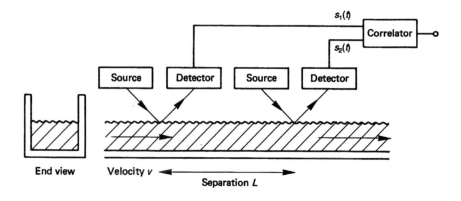

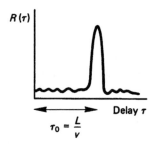

Figure 8.19 *Correlation flow-measurement system.*

produced alternately from the top and bottom surfaces and are known as the *von Karman vortex street*. The frequency of vortex shedding f is related to the fluid velocity v by $f = Sv/d$, where S is a constant known as the Strouhal number and d is the width of the body.

The vortices are regions of almost sinusoidal changes in fluid velocity and pressure, and may be detected by piezoelectric, thermal or ultrasonic methods. The vortex flow technique has the particular advantage that it provides a frequency (digital) output, like a turbine, but has no moving parts. It can have an accuracy of about 1 per cent for values of Reynolds number greater than 10^4.

8.4.2 Vector flow transducers

Apart from various forms of probes and vanes for measuring the direction of flow, the best-known methods are by means of Pitot tubes and anemometers. A Pitot tube is shown in figure 8.20. It involves a hollow tube feeding an inclined manometer, and has to be oriented to give a maximal reading. Pitot tubes are often used for volume flow measurement in pipes, especially when the profile across a section is needed.

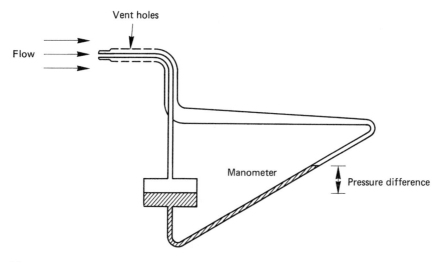

Figure 8.20 *Pitot tube flow transducer.*

The hot-wire anemometer comprises an electrically heated wire placed in the fluid. The amount of cooling depends on the direction and rate of flow and may be detected by a temperature transducer. Alternatively the wire may have a known resistance/temperature characteristic and can form one arm of a bridge circuit. The heating current may be maintained constant, but an alternative method is to employ a feedback system to maintain the wire at constant temperature, so the heating current becomes proportional to the flow. A schematic arrangement is shown in figure 8.21.

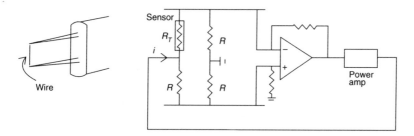

Figure 8.21 *Hot wire anemometer and control system.*

Any change in the temperature of the wire produces a change in its resistance, which is detected by the bridge circuit and compensated for by adjusting the heating current. Both wire and deposited film sensors are available.

8.5 Microphones

Microphones are essentially pressure transducers, but they differ from conventional pressure transducers in that they are intended to measure rapid changes in pressure, often at low amplitudes. The average limit of human audibility at a frequency of 1 kHz, for example, is 20 μPa r.m.s. and the noise in sound measuring systems can approach this level. Although many different transducer types are used in microphones, there are two main families. These are displacement types for microphones which respond to pressure, and velocity types for microphones which respond to pressure gradient. The pressure or pressure gradient is detected by a diaphragm or an elastically suspended disc or cone. In the case of pressure microphones, the back of the diaphragm is sealed from the surrounding air except for a minute hole which allows long-term pressure equalisation to occur. The pressure on the back of the diaphragm thus remains at the average surrounding air pressure. Short-term pressure changes on the front of the diaphragm therefore cause it to deflect. In the case of pressure-gradient microphones, the back of the diaphragm is open and it is the difference in frontal and rear pressures that causes a deflection of the diaphragm.

Pressure microphones are not directional in their response and are thus termed omnidirectional; pressure-gradient types have a cosine polar response and are thus termed bidirectional or figure-of-eight. Microphones that combine pressure and pressure gradient give a cardioid polar response and are usually pressure-gradient types with an acoustic phase-shifter or delay between the surrounding air and the rear of the diaphragm.

The pressure of a sound wave for any given energy flow is independent of frequency; thus the diaphragm of a small-pressure responding microphone is subjected to a force that is independent of frequency. The displacement of the diaphragm is thus independent of frequency, and a displacement transducer may be used to give an electrical output that is also independent of frequency. This is true for frequencies for which the wavelength is large compared to the size of the microphone. At these frequencies the sound diffracts around the microphone while at higher frequencies the microphone causes shadowing. The pressure on the diaphragm at high frequencies is thus dependent on the size and shape of the microphone (typical microphones are about 13 mm in diameter, which is equivalent to the wavelength of sound in air at 25 kHz).

The force applied to the diaphragm of a pressure-gradient microphone is, at low frequencies, proportional to frequency. This increase in force with frequency occurs until half the wavelength of the sound is equal to the effective thickness of the microphone. Above this frequency the pressure gradient (and hence the force applied to the diaphragm) falls, becoming zero when the wavelength of the sound is equal to the effective thickness. The size of the microphone also affects the diffraction of the sound, as it does for the pressure microphone. In addition pressure-gradient microphones are affected by the proximity effect. The pressure gradient rises by 6 dB/octave below a corner frequency, depending on the curvature of the wavefront. The frequency response of the microphone thus depends on the distance of the microphone from the sound source. Pressure-gradient microphones use velocity transducers to achieve an electrical response which is independent of frequency at high microphone-to-source separations. Switchable equalisers are usually incorporated into the microphones to attempt to compensate for the proximity effect.

8.5.1 Measurement microphones

Measurement microphones are almost invariably pressure types as the proximity effect makes the use of pressure-gradient types difficult. Although piezoelectric transducers are used in hydrophones (underwater microphones) where the high density of the medium matches the acoustic impedance of the transducer well, the majority of measurement microphones use variable-capacitance transducers. Using the diaphragm as one plate of the capacitor keeps the mass of the moving parts low and the construction simple. This results in a microphone of high responsivity and with minimal unwanted resonances. A backplate behind the diaphragm provides the other plate of the capacitor and, by the addition of holes or slots in the backplate, also provides acoustic damping of the diaphragm resonance. In a typical measuring microphone (figure 8.22) the diaphragm-to-backplate spacing might be about 25 μm. The diaphragm itself consists of a highly tensioned conductive film, either of pure metal, or of metallised plastic. The use of pure metal allows much higher tensions and gives the diaphragm immunity from attack by organic solvents. The fundamental resonance of the diaphragm is usually at the high-frequency limit of the microphone's frequency response (typically 10–40 kHz, although it may extend to 150 kHz for specialised miniature microphones). The diaphragm is thus compliance-controlled over most of its operating frequency range, but becomes resistance-controlled at high frequencies. The resonance is usually about critically damped by the holes in the backplate.

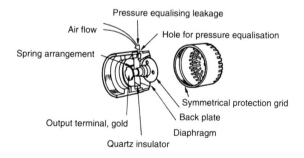

Figure 8.22 *Measurement microphone (courtesy of Brüel & Kjaer).*

8.5.2 *Microphone excitation methods and pre-amplifiers*

Capacitor microphones use two main types of excitation. These are a.c. excitation, using methods already described in chapters 4 and 5, and d.c. excitation either by a polarising voltage or by an electret coating on the backplate. The use of d.c. excitation has the advantage that an output is obtained directly, but the disadvantages that there is a low frequency limit to the frequency response and that audio frequency electrical noise is not rejected. This is a particular problem given that the input impedance of the pre-amplifier has to be of the order of 10 GΩ to achieve a satisfactory low-frequency response with a source capacitance of typically 17 pF. Low noise FET amplifier designs are used which achieve high input impedance by means of positive feedback (sometimes called bootstrapping). The pre-amplifier is mounted as close as possible to the microphone, usually in the same casing, and often incorporates double screening techniques to reduce stray capacitance and capacitive pick-up of noise.

The two-plate variable separation capacitive transducer is inherently non-linear, but the very small diaphragm displacements, typically 0.5 pm r.m.s. at 0 dB sound pressure level (in dB relative to 20 μPa r.m.s.), mean that the output is effectively linear up to about 150 dB SPL. These small diaphragm displacements, together with typical responsivities of around 10 mV/Pa, mean that the pre-amplifier noise is equivalent to a sound pressure level of about 20 dB (A-weighted). This is not usually a problem in practice, as background acoustic noise is rarely below this level.

8.6 Exercises

8.6.1. Use the general transfer function of an accelerometer (equations (8.1) and (8.2)) to deduce the types of instrument required for the following measurements:
  (i)   absolute velocity over the range 10 Hz to 1 kHz
  (ii) absolute acceleration over the range 10 Hz to 10 kHz
  (iii) absolute displacement over the range 0 to 100 Hz.

8.6.2. Figure 8.23 shows a strain gauge accelerometer, which consists of a mass of 0.1 kg attached to a beam of mild steel of elastic modulus $E = 2 \times 10^{11}$ N/m^2. A strain gauge is attached to each side of the beam and connected differentially in the bridge circuit of figure 8.23(b). Each gauge has an unstrained resistance of 1 kΩ and a gauge factor of 2.
  (i)   Find the natural undamped frequency of the device (assume that the mass of the beam is negligible).
  (ii) Find the zero-frequency acceleration responsivity at the amplifier output (assume that the beam is critically damped).
  (iii) Find the maximum input acceleration, assuming this is set by the amplifier limiting when its output reaches ±10 V.

Note: A force f applied to the end of the beam produces a deflection at the free end of $4fl^3/(Ebd^3)$ and a surface stress at distance x from the free end of $6fx/(bd^2)$.

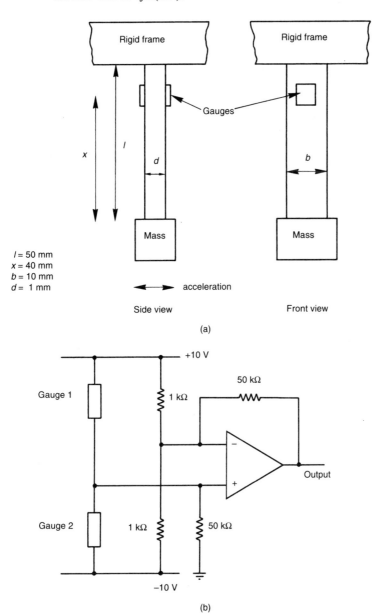

Figure 8.23 *Strain gauge accelerometer (a) and amplifier (b).*

8.6.3.   A d.c. force measurement system is required for forces up to 1000 N
using a cantilever beam clamped at one end with the force applied to the
free end. The cantilever is of mild steel and of length 5 cm, width 2 cm
and thickness 4 mm. Discuss the advantages and disadvantages of the
transducers below, and recommend the most suitable type:
  (i)   strain gauges
  (ii) piezoelectric crystals
  (iii) variable-separation capacitive transducers.
  [Use the equations given in exercise 8.6.2.]

8.6.4.   Discuss the transducers required for the following measurements:
  (i)   the rate of flow of a transparent liquid through a transparent pipe by
        a non-contacting method (bubbles or other disturbances may be
        assumed to be present)
  (ii) the rate of flow of water through an open channel
  (iii) the flow velocity at various positions and depths in a river.

9 Actuators

So far in this book we have concentrated on sensors and transducers which allow some physical parameter to be represented in an electrical form. These are the input transducers mentioned in chapter 1. In this chapter we will consider output transducers which take electrical signals and produce a physical effect (radiant energy giving light, mechanical energy giving physical movement, etc.). These perform three useful functions:

(1) they allow a human operator to observe the output, for example a moving coil meter
(2) they allow the output to be used for feedback, for example a coil/magnet assembly in a force–balance system (see chapter 8)
(3) they can produce a physical excitation, for example a piezoelectric ultrasonic transmitter in a sonar system (see chapter 10).

9.1 Electromagnetic actuators

These are normally force transducers, in this case producing a force from an electrical signal. Two effects are exploited in these transducers: the force generated on a current-carrying conductor in a magnetic field (see chapter 3) and magnetic attraction or repulsion due to an electromagnet.

9.1.1 Moving coil actuators

The basic construction of all moving coil actuators is the same in that they consist of a coil situated in an air gap between the poles of a magnet.

Dynamic transducer

This is the simplest form of moving coil actuator. A ring magnet and suitable pole pieces produce an intense magnetic field across a small circular air gap. A coil of fine wire, sometimes wound on a non-conducting former, is located centrally in the air gap by a highly compliant suspension. Connections to the coil are usually made along this suspension. Figure 9.1 shows a typical arrangement.

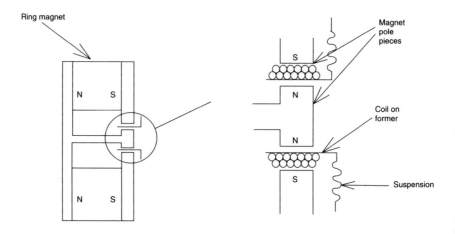

Figure 9.1 *Cross-section through coil–magnet assembly.*

In order to keep the force produced constant, but also allow movement of the coil, either the coil is made deeper than the air gap or vice versa. This means that the number of turns in the magnetic field does not change for small displacements; however, fringing of the field means that some non-linearity remains. A typical application of this transducer is the dynamic loudspeaker in which the coil is bonded on to a rigid cone. Movement of the cone produces sound waves over a limited frequency range and a number of differently sized cones are often used to cover the audio frequency range. (It should be noted that loudspeakers intended for rock music use have voice coils of about the same depth as the air gap. This makes them highly non-linear but able to dissipate large amounts of power before they are thermally destroyed.)

Strictly speaking, coil/magnet assemblies should be current driven as the force $F = (Bl)i$. In practice, they are often voltage driven with the assumption that the coil resistance is constant. As the coils are usually made of copper which like all metals has a temperature-dependent resistivity, the self-heating effect of the current introduces some non-linearity. This can be quite significant if high powers are involved and in these cases current drive is recommended.

d'Arsonval mechanism

Despite being invented over one hundred years ago, the ordinary d.c. ammeter is still in use today. Although it has widely been replaced by the digital meter, it still has some advantages. Being an analogue device it is much easier to take approximate readings or to watch trends. Some digital meters try to mimic this by adding a bar-graph display to the normal numeric display. Figure 9.2 shows a typical d.c. ammeter.

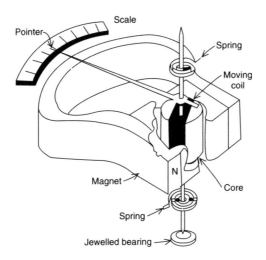

Figure 9.2 *D.c. ammeter.*

The coil is wound on a rectangular former and positioned between a central core and the poles of a magnet by jewelled bearings. Connections to the coil are made through two flat coil springs which also provide a rotational restoring force. Current flowing through the coil produces a turning torque which is resisted by the two springs. Displacement of the pointer is thus relatively linear over a limited angle. The fixed end of the top spring is usually attached to a screw on the front of the meter which allows the zero position to be set.

D.c. motor

The d.c. motor works on the same principle as the meter movement mentioned in the previous section. The coil or armature is usually wound on a soft-iron core which is located between the poles of the permanent magnet. A split-ring commutator and brushes are used to reverse the direction of the current flow at the point at which the coil moves out of the magnetic field. Figure 9.3 shows the basic principle of operation.

In practice, multi-pole motors are used with many segments on the commutator so as to ensure a more constant torque over the full 360° rotation. These have more than one coil, and commutation is arranged to switch as one coil leaves the magnetic field and the next one enters. Ironless rotor motors have a coil with no iron core but instead a fixed steel tube inside the coil to complete the

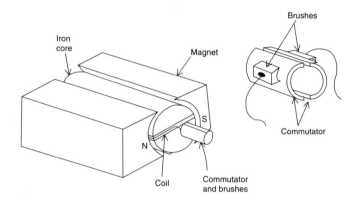

Figure 9.3 *D.c. motor.*

magnetic circuit. In this respect they are similar to the meter movement. They have much lower rotor mass and thus accelerate more quickly than conventional types and are often used in small servomechanisms. A typical ironless rotor d.c. motor is shown in figure 9.4.

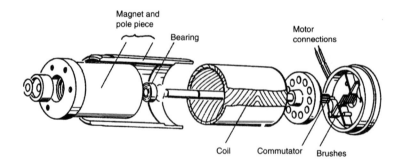

Figure 9.4 *Ironless rotor d.c. motor.*

The torque produced by a d.c. motor for a given current depends on the length of conductor in the magnetic field, the strength of the field and the diameter of the rotor. These factors are usually combined into a single constant k_m which also gives the speed-to-voltage relationship of the motor (see chapter 3):

$$\tau = k_{\mathrm{m}}I \qquad \omega = \frac{V}{k_{\mathrm{m}}}$$

Because of the action of the commutator and brushes, the value of k_{m} and of the armature resistance may vary with the angle of the armature in some motors. This effect is known as 'cogging' and in this respect multi-pole motors, particularly ironless rotor types, are superior.

9.1.2 Moving iron actuators

These can be split into two basic types: those that are coil/magnet assemblies with fixed coils and those that rely on the magnetic force generated when a soft-iron core is magnetised by a current flowing through a coil. Those in the first category differ little from moving coil types and produce a force or torque proportional to current. Those using magnetic attraction or repulsion, however, are usually highly non-linear.

Figure 9.5 *Basic principle of stepper motor.*

Stepper motor

The rotor of a stepper motor consists of a permanent magnet or series of permanent magnets. The stator windings are wound on soft-iron cores which surround the rotor as shown in figure 9.5.

If the two horizontal windings are energised as shown, the rotor will attempt to align itself horizontally. Energising the vertical windings would similarly cause the rotor to align itself vertically. Intermediate angles may be obtained by partially energising both windings. Real stepper motors have many windings and can achieve small step angles (typically $< 2°$) when driven with the appropriate signals. Figure 9.6 shows a precision stepper motor with only two of the stator windings shown for clarity.

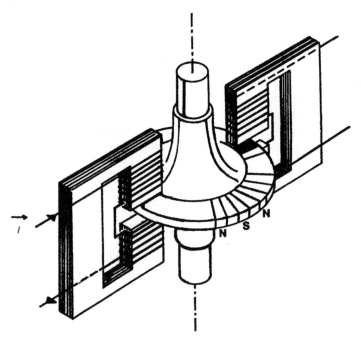

Figure 9.6 *Typical precision stepper motor.*

Stepper motors are useful in positioning light loads from computer control. Their inherent positional feedback operates only over a very small angle and excessive loads may cause steps to be missed or the rotor to overshoot to an incorrect position. Stepper motors are much loved by computer programmers who believe that the rotor will go to precisely where they have told it. True position control can only be achieved, however, by the use of feedback from some form of angular or rotational transducer (see chapter 5).

Brushless d.c. motor

The brushless d.c. motor is a special kind of stepper motor in which the stator is energised, depending on the position of the rotor. This is measured by Hall-effect devices which are placed around the permanent magnet rotor or a small reference permanent magnet. External circuitry is then used to switch the appropriate stator windings so as to achieve constant rotation. The advantages of brushless d.c. motors are that there are no brushes to wear and that electromagnetic interference may be reduced. A typical brushless d.c. motor is shown in figure 9.7.

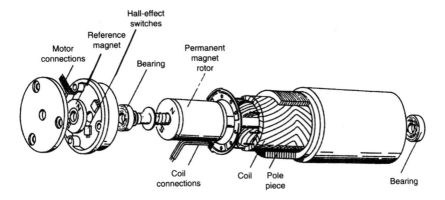

Figure 9.7 *Brushless d.c. motor.*

Solenoids

Strictly speaking, the word 'solenoid' refers to a long cylindrical coil of wire which may act as a magnet when carrying a current. The word has come to refer also to the mechanical actuator which uses a solenoid to produce a magnetic field and hence apply a force to a soft-iron core. Initially the core is only partially within the coil, as shown in figure 9.8.

Once a current flows in the coil, a force is exerted on the core so as to make it move in a direction that increases the inductance of the coil. The core thus tries to centralise itself vertically within the coil. The force f is given by

$$f = \frac{1}{2}I^2 \frac{dL}{dx}$$

where L is the inductance of the coil and x is the displacement of the core. This relationship is valid until the core becomes saturated when the available force will be less than predicted. Figure 9.9 shows a set of force versus displacement graphs for a typical solenoid at different currents.

Because of its inherent non-linearity the solenoid is of limited use as an

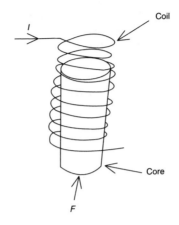

Figure 9.8 *Typical solenoid.*

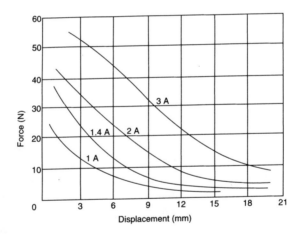

Figure 9.9 *Force versus displacement of typical solenoid at different coil currents.*

actuator for measurement purposes, but is widely used to provide a mechanical switching action from an electrical signal.

Moving iron meter

The moving iron meter, unlike the moving coil meter, is only used in particular circumstances. The reason is that, like the solenoid, it is highly non-linear. This does not mean that it is not capable of providing an accurate measure of current but that the range is limited. The non-linearity of the meter is compensated in the scale which is bunched at the ends and spread out in the middle. The basic principle of operation is shown in figure 9.10.

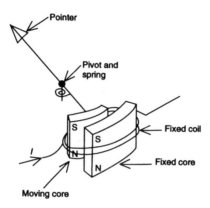

Figure 9.10 *Moving iron meter.*

A current I flowing in the coil magnetises the two soft iron cores. As the field is the same for both, they will be magnetised with the same polarity and hence repel each other. This will be true irrespective of the direction of the current I and hence the meter will respond to a.c. currents as well as d.c. currents. This type of meter is thus very useful for monitoring a.c. power supplies, as the extended centre scale allows small changes from the normal value to be easily read. The fixed coil is also more rugged than a delicate moving coil, and the meter can therefore be used in more hostile environments, such as on mobile electrical generators.

Relays

The relay is an electrically activated switch which gives a high degree of electrical isolation between the control and switched circuits. The switch contacts are attached to a moving armature which is attracted to the core of a coil as shown in figure 9.11.

Similar devices without the contacts were used to give an acoustical output in early telegraph systems. Two distinct sounds were made by the relay opening and closing, and hence the difference between a long and short signal could be heard.

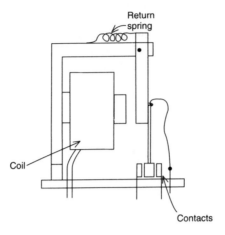

Figure 9.11   *Typical relay.*

9.2   Electrostatic actuators

If the plates of a parallel plate capacitor are charged to a voltage V, then there will be a force F attracting the two plates together; where F is given by

$$F = \frac{\varepsilon_0 A V^2}{2d^2}$$

where A is the area of the plates and d is the separation. This force is quite small unless the separation is small or the voltage large. For example, the attractive force between two plates each of area 1 cm^2, separated by a distance of 1 mm and with a voltage of 10 V across them is only 4.4×10^{-8} newtons.

Electrostatic actuators are generally of limited use because of the high voltages required for reasonable plate separations but there are two significant exceptions. If the size of the actuator is very small indeed, for example it is integrated in silicon, then the forces become useful (if all the lengths in the previous example are reduced by the same factor, there is no change in force). At these scales, coils are difficult to manufacture but the simple structure of the electrostatic stepper motor has allowed it to be integrated in silicon to make microscopic rotational actuators. Figure 9.12 shows the basic principle of the electrostatic stepper motor (the drive electronics would be integrated on to the same chip, but are not shown here).

The other significant advantage that the electrostatic force transducer has to offer, is that a force may be applied to a conducting surface without requiring any mechanical connection. This means, for example, that a diaphragm of very low

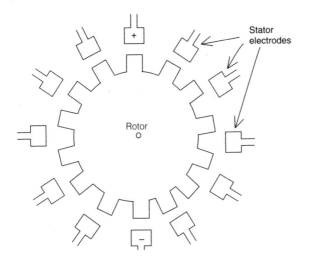

Figure 9.12 *Electrostatic stepper motor.*

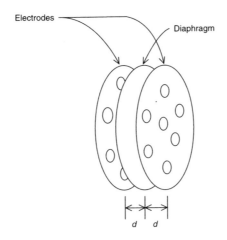

Figure 9.13 *Diaphragm flanked by two electrodes.*

mass may be moved without requiring any further mass to be added to it. This characteristic of electrostatic actuators has been used to advantage in the production of the electrostatic loudspeaker which, because of the low mass of its diaphragm, can produce very fast transients accurately. A modified type of actuator is required, however, in order to keep the relationship between force and voltage linear, as it is normally a square law. Figure 9.13 shows a diaphragm flanked by two electrodes.

If the voltages on the electrodes are made $+V_p$ and $-V_p$ then the force on the diaphragm F becomes

$$F = \frac{2\varepsilon_0 A V_p V_s}{d^2}$$

which is linear with respect to the voltage V_s but will vary if the diaphragm actually moves. Displacement of the diaphragm must therefore be kept small compared to the diaphragm-to-electrode spacing if non-linearity is to be kept to a low level.

A force feedback (force–balance) pressure transducer or microphone can be constructed if the plates of the capacitor are also made to function as a variable-separation capacitive transducer using high-frequency excitation. The output of the capacitive transducer (once demodulated) is amplified and fed back to the diaphragm in antiphase. Any pressure that produces a force acting on the diaphragm is thus resisted by the electrostatic feedback force. The output of the system is the voltage on the diaphragm as this produces the balancing force, which in turn cancels the input force. The responsivity of such a device no longer depends on the characteristics of the diaphragm but only on the area of the electrodes, the diaphragm-to-electrode spacing and the polarising voltage. As the function of such a device is to attempt to hold the diaphragm still, the spacings do not change significantly and the device is highly linear. The resonant characteristics of the diaphragm are reduced by the loop gain of the system and thus a flat frequency response is easily obtained over a wide bandwidth.

9.3   Electro-optic devices

There are a great many electro-optic devices which may be used to provide a numeric or graphical display of the output of a measurement system. There are three types that are particularly common: the light emitting diode (LED), the liquid crystal display (LCD) and the cathode ray tube (CRT). It is beyond the scope of this book to describe the function of the LCD and CRT, as even these common types are complex in their operation. We will mention the LED, however, as it also finds use as a source to excite optical systems.

9.3.1 The LED

If a *p–n* diode is forward-biased then the normally empty electron states in the conduction band of the *p*-type material and the normally empty hole states in the valence band of the *n*-type material become populated by the injected carriers. These carriers recombine across the band-gap, releasing energy approximately equal to the band-gap energy. This energy may be radiated or dissipated as heat and the ratio of radiated to non-radiated energy depends on the materials used. If the energy is radiated then a photon is created with a frequency given by

$$E_g \approx h\nu$$

where E_g is the band-gap energy, ν is the frequency and h is Planck's constant.

There are two types of band-gap in semiconductors: direct band-gaps where the electrons and holes have the same momentum, and indirect band-gaps where the momenta differ. Direct band-gap materials are the most useful as the electrons and holes may easily recombine, giving off a photon. A commonly used direct band-gap material is gallium arsenide (GaAs). In indirect band-gap materials, the electron must change momentum before it can recombine with a hole. To change this momentum, a third particle called a phonon must be emitted or absorbed. This three particle (electron, photon and phonon) process is far less likely than the simpler two particle process (electron and photon) and so most of the energy is lost as heat rather than creating a photon. Common indirect band-gap materials which give negligibly low levels of electroluminescence are silicon and germanium. If the difference in momenta in the indirect band-gap material is small, impurities may be added to increase the probability of radiative recombination greatly. Thus gallium phosphide may become a useful material if nitrogen is added, but both silicon and germanium have too large a momentum difference for this to be possible.

The simplest structure for an LED is the planar structure shown in figure 9.14.

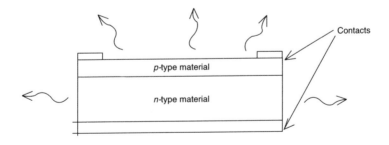

Figure 9.14 *Structure of planar LED.*

The distribution of light with angle from the surface of a planar LED is Lambertian and much of the light is internally reflected at the crystal–air interface. This means that the overall efficiency (optical power out/electrical power in) of the LED is low, typically around 1 per cent. The relationship between current and light output is essentially linear and thus amplitude modulation of the light is readily achieved. LEDs are available in the wavelength range 0.5–1.5 μm approximately, with the most common being in the range 0.6–0.9 μm. Specially fabricated types are available to couple efficiently with optical fibres, and types designed specifically for communications may be modulated at hundreds of MHz. More common types have slower response times and may be modulated at between 500 kHz and about 10 MHz, depending on their application. This makes them well suited to use in a.c.-excited optical measurement systems.

9.3.2   The laser diode

The laser diode is basically a special type of LED that contains an optical Fabry–Perot cavity. The term LASER is an acronym for Light Amplification by Stimulated Emission of Radiation and the basic laser is thus an amplifier rather than a source. Most lasers also contain a source and act as optical oscillators. In the case of the laser diode, the source functions in the same way as a conventional LED by spontaneous emission. Photons may also be created by stimulation however; that is, one photon interacting with an electron in a high-energy state causes it to fall to a lower state, creating another photon of identical wavelength and phase. Above a certain value of stimulation, the creation of photons exceeds the absorption in the cavity and lasing occurs. In the laser diode, the ends of the junction are cleaved or polished to form mirrors, and complex structures are used to create an optical waveguide which forms the sides of the cavity.

The relationship between light output and current in a typical laser diode is shown in figure 9.15.

Below the threshold current, the device functions in the same way as a

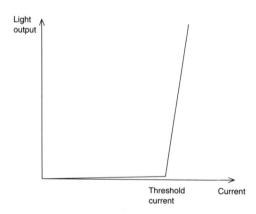

Figure 9.15   *Light output versus current for typical laser diode.*

standard LED by spontaneous emission but, above the threshold current, lasing occurs and the efficiency is much increased. The output also becomes coherent, although the coherence length is short by comparison with gas lasers (typically only 10 mm). The laser diode is thus suitable as a source for interferometric measurements either in air or through mono-mode optical fibres.

9.4 Piezoelectric actuators

It was mentioned in chapter 3 that the piezoelectric effect was reversible, a displacement resulting from a change in charge. This effect may be exploited to produce an actuator whereby the application of a voltage across a disc of piezo-electric material (such as quartz) will cause it to change thickness. The relationship is

$$x = dv$$

where x is the displacement, v the applied voltage and d the 'd coefficient' of the material. It should be realised that the units of d given in chapter 3 (C/N) are equivalent to m/V. The typical value for lead zirconate titanate (PbZ) is thus 150×10^{-12} m/V. As this results in rather small displacements for realistic voltages, actuators using this effect tend to be used for micro-positioning (for example, the positioning of optical components).

Another use for the piezoelectric actuator is in the generation of acoustic waves in air or water. Piezoelectric actuators are often used to produce high-frequency sound in air although they are more efficient when used in denser mediums such as water. When used as an ultrasonic source, the crystal is usually made to resonate mechanically at the required frequency (typically 40 kHz) by the addition of a cone of suitable mass. This resonance increases the efficiency over a narrow frequency range but also results in a poor response to the tone bursts typically used in sonar systems.

9.5 Exercises

9.5.1. A coil–magnet assembly has a Bl factor of 10 newtons/amp. If the coil has a resistance of 7 Ω and a maximum power dissipation of 50 W, what is the maximum available force?

9.5.2. A d'Arsonval meter movement has two restoring springs, each of strength 2.5×10^{-7} N m/radian. If the full-scale deflection of the meter is 2 radians and the coil–magnet produces a torque of 0.01 N m/A, what is the full-scale deflection current?

9.5.3. A d.c. motor gives a no-load speed of 15 000 revolutions per minute for an armature voltage of 6 V. If the maximum permissible armature current is 3 A, what is the maximum torque of the motor?

9.5.4. A simple electrostatic actuator has two parallel plates each of area 50 mm^2 separated by an air gap of 10×10^{-3} mm. If the maximum possible voltage across the plates is 50 V, what is the maximum available attractive force?

10 Measurement Systems

In chapters 5 to 8 we dealt with transducers from the functional point of view, discussing the devices available for the measurement of length, temperature, light, acceleration, flow, etc. In this chapter we will consider various important measurement systems, including solid-state sensors, resonator sensors, optical fibre systems, pyrometry and ultrasonics, where a particular technology or technique is capable of the measurement of several different quantities, for example displacement, speed and flow all by ultrasonics. We will give a review of each technology first, followed by a discussion of applications.

10.1 Solid-state transducers

Most solid-state transducers are based on silicon. It has been said that silicon is not an ideal material for sensors, but nevertheless it has a number of excellent properties. The most important are the ease with which it can be purified to almost unbelievable levels and the fact that its oxide, produced by heating in an oxygen atmosphere, forms an excellent mask for the selective doping and etching of a silicon-based structure. Most of our present technology depends on these properties, of course, but similar possibilities clearly exist for the development of transducing devices employing essentially the same fabrication techniques.

Table 10.1 shows the various forms of energy again, with the known modulating and self-generating transducer effects. The only omissions are mechanical and magnetic self-generators, silicon being neither piezoelectric nor ferromagnetic. However, both these defects can be overcome by the deposition of

Table 10.1 Transduction effects in silicon

Energy type	Self-generating	Modulating
Mechanical	–	Piezoresistive
Thermal	Seebeck	Resistance/temperature
Radiant	Photovoltaic	Photoconductive
Magnetic	–	Magnetoresistive, Hall
Chemical	Galvanoelectric	Ion-sensitive

suitable layers – for example, of piezoelectric ZnO and ferromagnetic NiFe. In fact, the ease of deposition of various materials is a particularly powerful feature and has already led to many important new devices, several being in the chemical field. We will discuss some of the important transducers, classifying them by energy type in this case.

10.1.1   Mechanical transducers

Pressure transducers using the piezoresistive effect are the most widely used devices. As explained above, metallic strain gauges have a gauge factor of about 2, purely owing to dimensional effects, but this is far outweighed by the piezoresistive effect in semiconductor gauges, with factors of several hundred. Some pressure transducers are now integral devices, a small (1 mm diameter) diaphragm being etched from n-type silicon and with four p-type resistors diffused on to it in a bridge arrangement. Such devices are available commercially and are widely used, though active temperature compensation is necessary.

Other physical effects usable in pressure transducers are capacitance changes and changes in the characteristics of planar transistors with pressure applied to the emitter surface. MOSFETs are also pressure sensitive and this can be increased by a layer of piezoelectric ZnO.

An accelerometer has been developed recently, comprising etched silicon structures in the form of a minute silicon mass (1 mm^2 area and 200 μm thick) connected to a 0.5 mm silicon cantilever beam, which includes a diffused silicon piezoresistor. It has an acceleration range of 0.1–1000 m s^{-2}. The fabrication of such a three-dimensional structure requires very precise control of the etching processes, but the excellent properties of silicon have permitted the development of various devices of this form. The piezoresistive effect has also been used in anemometers, in which differential cooling of diffused silicon resistors in a bridge arrangement gives a measure of the rate of flow of air.

10.1.2   Thermal transducers

The band-gap in silicon is comparatively large, so doping is necessary to produce useful temperature effects (that is, silicon behaves as an extrinsic semiconductor at room temperature). A few resistance thermometers have been developed, mostly for rather specialised applications. However, the most widely used effect is the temperature-dependent voltage across a forward-biased diode, discussed above. The linearity and stability are not high, but the possibility exists of adding some signal-processing elements on the same chip and much development is taking place in this direction. Another approach is to use two transistors, usually a dual transistor, operated at a fixed emitter/current ratio, when it is found that the difference in base–emitter voltages is proportional to absolute temperature. Again it is necessary to include some processing circuitry and a compact accurate device can be fabricated.

10.1.3 Radiation transducers

Photoconductors and photodiodes are the best-known examples of silicon radiation transducers, and have been described in some detail in chapter 7. Charge-coupled devices (CCDs) are now widely used in solid-state television cameras. They utilise the dependence of capacitance on voltage in MOS diodes, and consist of an array of diodes in the form of a shift register. Light falling on the device generates packets of charge, which can be manipulated by suitable applied voltages. Parallel arrangements of large numbers of such registers enable two-dimensional images to be analysed and arrays of up to a million elements have been produced.

10.1.4 Magnetic transducers

Since silicon is not magnetic, only modulating magnetic transducers are available; these mostly employ either the Hall effect or the magnetoresistive effect discussed in chapter 3. Silicon has a high Hall coefficient, and thin Hall plates are available comprising resistive *n*-type layers on a *p*-type substrate. They may be used for the measurement of fairly large magnetic fields or electric current (with a fixed magnetic field) and often include some signal processing on the same chip.

The magnetoresistive effect in silicon is fairly small, so silicon is usually used as a substrate for more suitable materials, such as indium antinomide. Films of ferromagnetic NiFe are also used, and can exhibit significant changes in resistance due to rotation of the direction of magnetisation.

The characteristics of diodes and transistors show a dependence on magnetic force. The most suitable arrangements comprise some form of dual transistor, the difference in collector currents being proportional to field strength. The devices are also sensitive to field direction.

10.1.5 Chemical transducers

In the past, most sensors for chemical quantities have tended to be rather bulky and complicated, but the area has received a large impetus with the development of various miniature solid-state sensors. The first device was the ISFET (ion-sensitive FET) comprising a MOSFET with an ion-selective layer of a polymer (Si_3N_4, Al_2O_3) grown on the gate. The characteristics are affected when immersed in an electrolyte, and the pH can be measured. The palladium-gate

MOSFET was later developed, the replacement of the normal gate by palladium leading to a sensitivity to hydrogen gas.

A range of transducers has been produced in which a suitable polymer is deposited on an electrode structure or on the gate of silicon MOSFETs. Devices sensitive to CO, CO_2, CH_4, SO_2 and NH_3 have been made. They are usually sensitive to moisture content as well, and various humidity sensors are available. Another form of humidity sensor involves an electrode structure for detecting capacitance changes caused by dew formation, together with a semiconductor temperature sensor and miniature Peltier cooling element. It is really a miniature version of a conventional humidity sensor, using silicon as the substrate. A similar technique is used in an oxygen sensor, using etched silicon channels containing an electrolyte of NaCl, the cell current being proportional to the pressure of oxygen diffusing through a suitable membrane.

10.1.6   Other materials

Various other materials have been used for solid-state sensors, notably germanium and gallium arsenide. The latter is considered to have great potential, but does not have the advantage of silicon, whose oxide forms such an excellent mask, making large-scale integration less convenient.

A particularly promising material for piezoelectric and pyroelectric sensors is polyvinylidene fluoride (PVDF). Most piezoelectric materials have very high stiffness, limiting their applications to pressure and force transducers, but PVDF is available in thin films of low stiffness and also relatively low permittivity, and will find considerable application in the next few years.

A review of silicon transducers has been given by Middlehoek and Noorlag (1981a).

10.2   Resonator sensors

Resonator sensors are mechanical modifiers in which a mechanical element is excited into vibration at its natural frequency, the value of which depends on the desired input quantity. The output is thus at a frequency proportional to the quantity of interest.

The advantages usually claimed of a frequency output, as opposed to an analogue voltage, are ease of interfacing to digital processors, stability, freedom from electrical interference and low power requirements. Actually the main advantage of a frequency output is that of ease of transmission, which is not very relevant here, and interfacing still requires a counting/timing system corresponding to an analogue-to-digital converter for an analogue signal, though digitisation of a frequency or time signal is easier and usually much more accurate. There is, of course, a cost associated with the conversion to frequency: the resonance is often very dependent on temperature and is rarely linearly proportional to the quantity required. The physical process being employed is, of course, perfectly analogue – it is just being used in a different way.

Many physical parameters may be measured by resonator sensor methods, usually using a wire, beam or cylinder. Vibrating-wire devices for measuring force or strain were the first to be developed, and consist of a stretched wire as in figure 10.1.

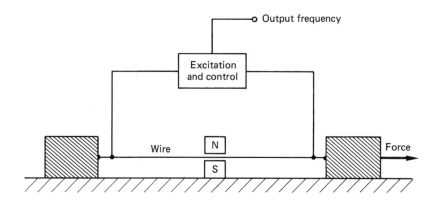

Figure 10.1 *Vibrating-wire force transducer.*

The resonant frequency f_r of a wire of length l and mass per unit length m, stretched by a force T is

$$f_r = \frac{1}{2l} \sqrt{\left(\frac{T}{m}\right)}$$

Wires of tungsten or indium are usually best. Excitation is by means of a large permanent magnet and a driver circuit which feeds current through the wire, and maintains the frequency at resonance by sensing changes in the electrical impedance of the vibrating wire. Some compact and accurate pressure transducers have been developed recently in which the wire is attached to a diaphragm, though considerable processing is required to remove temperature effects and produce a linear output.

Vibrating beams follow similar principles to vibrating wires, being used mostly for force, though other parameters can be measured. Figure 10.2 shows a beam used in flexure for level measurement, a similar arrangement for flow, a beam vibrating along its length for viscosity measurement, and a beam in torsional vibration for density measurement.

Resonating beams for force measurement are often in the form of two tuning forks connected tine-to-tine, as shown in figure 10.3, providing very high Q-factors. A tuning fork arrangement has a considerable advantage over a simple

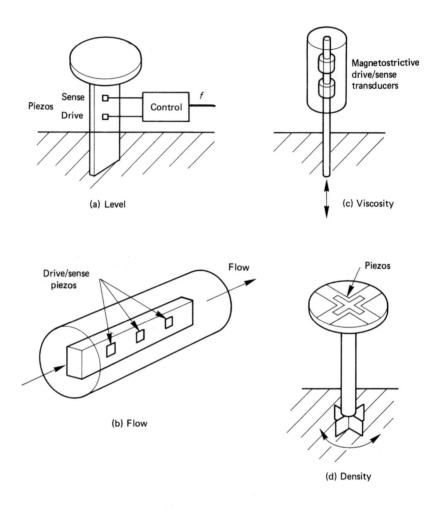

Figure 10.2 *Vibrating-beam transducers.*

beam, in that the energy loss at the clamping points is very small, and they are known as DETF (double-ended tuning fork) devices. They may be made very small, and some of the most recent devices are of quartz or even etched from silicon, in which case the dimensions may be less than 1 mm. The mechanical design is very important since it must be possible to excite a single vibratory mode of high Q-factor, and thermal effects must be minimised.

Vibrating cylinders and tubes are also popular. Excitation of a suitable mode is more difficult and substantial processing is required to obtain a useful output, but several different quantities may be measured in this way. Figure 10.4 shows a

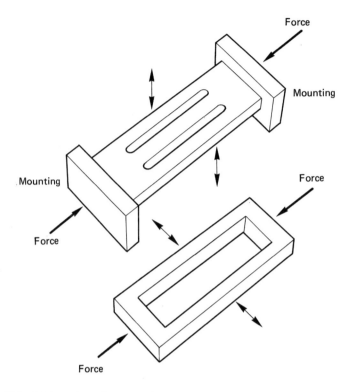

Figure 10.3 *Miniature beam force transducers.*

'tuning fork' arrangement for density measurement and a tube in torsional vibration for detecting mass flow.

Several sensors for chemical quantities are now available. They mostly use a small quartz crystal with a suitable thin film deposited on one face, as in figure 10.5, the resonant frequency changing as molecules are absorbed by the film. Moisture, or specific chemicals, can be detected with considerable accuracy. Usually two crystals are mounted side by side, one having no film and being used for reference purposes. Mass changes at the microgramme level can be detected.

10.3 Optical fibre transducers

The considerable developments in optical fibres over the last decade were primarily aimed at the communications field and it is only in the last few years that the possibilities for transducer development have become apparent. An optical fibre is a thin low-loss structure of glass or plastic in which the total internal reflection of a beam of light entering at one end causes the beam to be

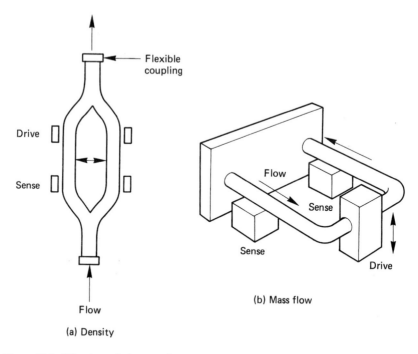

(a) Density

(b) Mass flow

Figure 10.4   *Vibrating-cylinder transducers.*

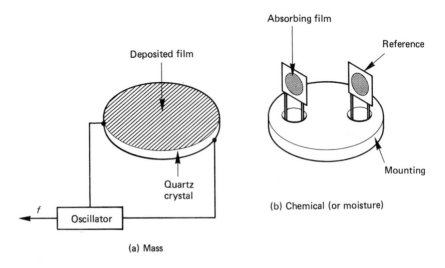

(a) Mass

(b) Chemical (or moisture)

Figure 10.5   *Mass and chemical sensors.*

contained completely within the fibre.

Figure 10.6 illustrates Snell's law of refraction of light at a boundary between two media of refractive indices n_1 and n_2 and the critical angle of incidence θ_c occurs for $\sin\theta_1 \geq n_2/n_1$.

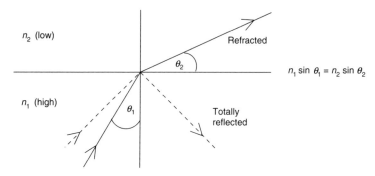

Figure 10.6 *Snell's law of refraction.*

Rays at angles of θ_c or greater are totally reflected and contained within the medium of higher index. Practical fibres always consist of a core of high index material surrounded by a cladding of lower index, as shown in figure 10.7, because the electromagnetic wave extends slightly into the low index material and would otherwise suffer serious attenuation if the fibre was handled or contacted in any way. The maximum entry angle φ_m of a ray to the fibre such that it becomes trapped within it (that is, $\theta_1 \geq \theta_c$) is given by $\sin\varphi_m = (n_1^2 - n_2^2)$ when the outer medium is air of index $n_0 = 1$. $\sin\varphi_m$ is called the numerical aperture (NA).

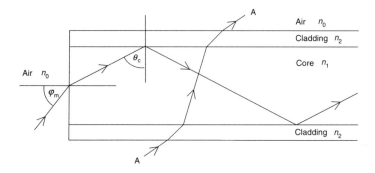

Figure 10.7 *Rays in a step-index fibre.*

Writing $n_2 = n_1(1-\delta)$ where $\delta \approx 0.05$, NA $\approx n_1 \sqrt{(2\delta)}$ and has a value of 0.47 with $n_1 = 1.5$, giving $\varphi_m = 28°$. It is clear that a ray, such as A, striking the fibre along its length cannot become trapped within it, and this freedom from interference or corruption is one of the big advantages of optical fibre systems.

The above type of fibre is known as step index, in contrast to graded index fibres in which there is a gradual change in index beyond a central core region. Such fibres have advantages in communications but fibres for sensing are always step index. There are two types of step index fibre, mono-mode fibres with a very thin core of diameter about 10 μm surrounded by cladding of thickness about 100 μm, and multi-mode fibres with cores of 50–100 μm and cladding 100–200 μm. Mono-mode fibres can transmit only one electromagnetic mode of vibration, corresponding to a ray that goes down the centre of the fibre with no reflections. In multi-mode fibres, the different possible angles of rays within the fibre correspond to different spatial patterns (modes) of the electromagnetic wave.

In addition to the glass fibres described above, plastic multi-mode step index fibres are also available. These have greatly inferior optical characteristics but often have significant advantages for sensing purposes, being easier to cut and handle. They usually have a core of 1 mm diameter surrounded by a thin cladding layer and a protective plastic sheath.

LED sources are used for multi-mode fibres, often in the near infra-red region, the light being detected with silicon phototransistors or photodiodes. Specialised optical couplings are available for attachment to the LEDs and detectors, for both plastic and glass fibres. With mono-mode fibres it is necessary to use a laser diode source, and such fibres are used for the more specialised measurement applications for which coherent light is required.

There are two distinct ways in which optical fibres may be used for sensing physical (or chemical) quantities. They can be used simply as signal guides to and from the area of interest, such as an oven in a temperature-measurement system, or a moving vane in a displacement system. Alternatively, the light passing through the fibre can be affected in some way by an external parameter; for example, the refractive index changes if the fibre is stressed. The particular advantages offered by optical fibres in transduction are their inherent immunity to electromagnetic interference and the ease with which they can be linked to an optical communication system.

The properties of light available for modulation in sensing applications are intensity, phase, polarisation, wavelength and spectral distribution. All these properties have already been used in sensors, though most devices use one of the first three. In terms of the six forms of energy discussed in chapter 1 we can find modulating effects in most cases, as shown in table 10.2, though of course there are no self-generators. However, in looking at examples of various transducers, it is easier to group them by the property of light that they employ.

10.3.1   *Intensity modulators*

Signal-guiding intensity modulators are the most widely used devices and some

Table 10.2   Modulating effects in optical fibres

Energy type	Modulating effects	Physical effect
Mechanical	Stress birefringence	Refractive index and absorption
	Piezo-absorption	Absorption
Electrical	Electro-optic effect	Refractive index
Magnetic	Magneto-optic	Refractive index
	Faraday effect	Polarisation
Temperature	Thermal	Refractive index and absorption

typical arrangements for displacement measurement are shown in figure 10.8 – (a) and (b) for transverse displacement, and (c) and (d) for translational displacement.

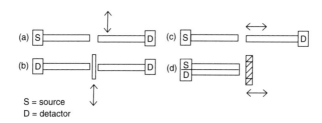

S = source
D = detactor

Figure 10.8   *Intensity modulators for displacement.*

It is less desirable to move the fibre, and arrangements (b) and (d) are preferred. The two fibre and mirror system in figure 10.8(d) is very widely used and figure 10.9 shows a plot of reflected light intensity versus displacement. The linear region is of the order of the fibre thickness and a working point must be chosen in the middle of the range.

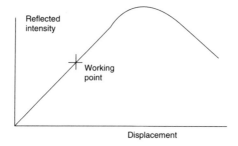

Figure 10.9   *Light intensity versus displacement in a reflective sensor.*

Differential three-fibre designs are possible and have the advantage that the output in the central position is zero and they can be used directly in control systems.

LEDs are used in the above systems and may be operated using d.c. or a.c. Operation with a.c., in which the light is modulated at, say, 10 kHz, is preferable from the point of view of noise, though a phase-sensitive detector is then required to produce a d.c. output.

Figure 10.10 shows an external-parameter intensity modulator, in which 'microbending' of the fibre is produced by an applied force, leading to a fall in transmitted intensity. Any bending of the fibre will cause the higher-order modes to escape, since they strike the fibre boundary at angles less than the critical angle, and the effect has been used in a 'data glove' for research in virtual reality.

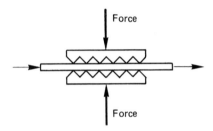

Figure 10.10   *Intensity modulator for force.*

10.3.2   *Phase modulators*

Phase changes can be produced by mechanical strain, temperature, etc., but the effect is mostly used for more rapidly changing parameters such as acoustic fields in a recently developed hydrophone. Interferometric methods are used, with a coherent source. One of the most common schemes is the Mach–Zender interferometer in figure 10.11(a). A mono-mode fibre is used with a modulator, usually a Bragg cell, which causes a suitable difference frequency to be produced, at which the phase shift is detected. A simpler method using a multi-mode fibre, in which phase modulation occurs between different modes in the same fibre, is shown in figure 10.11(b). An important application is the fibre-optic gyroscope; light is directed in opposite directions in a coil of fibre and a small phase shift is produced on rotation.

10.3.3   *Polarisation modulators*

Most fibres are internally birefringent, but unfortunately this varies with various conditions such as temperature, stress, etc. and also with time, so specially constructed fibres are necessary. The main application is in measuring large

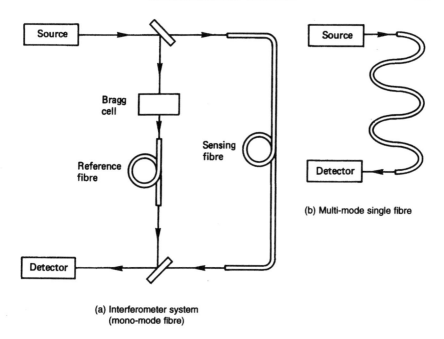

(a) Interferometer system
(mono-mode fibre)

(b) Multi-mode single fibre

Figure 10.11 *Phase modulators.*

electric currents by means of the Faraday rotation induced by a magnetic field, as in figure 10.12. The effect is proportional to the line integral of the field over the length of the fibre and, although small, has proved useful for high-voltage power lines, with a resolution of a few tens of amps.

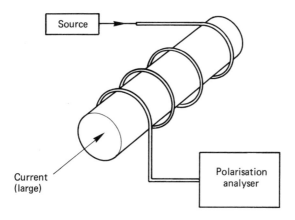

Figure 10.12 *Polarisation modulator for current measurement.*

Miniature temperature probes have recently been developed in which the temperature dependence of orthogonally polarised modes is exploited.

10.3.4 Wavelength modulators and spectral distribution modulators

These devices are mostly of the signal-guide type. The Doppler anemometer in figure 10.13(a) uses the Doppler shift of light scattered from moving particles. A number of temperature-measuring probes have been developed in which the light is guided to a special phosphor which emits spectral lines whose intensity ratio is temperature dependent; figure 10.13(b) shows a schematic arrangement.

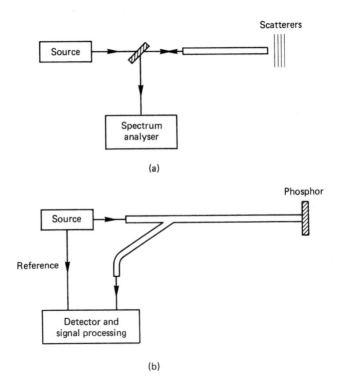

Figure 10.13 *Wavelength and spectral distribution modulators.*

This latter example could almost be described as a spectral distribution modulator, of which the best known is the optical-fibre pyrometer, which uses the same principle of black body distribution as in a normal pyrometer but avoids the unpredictable atmospheric effects that perturb these instruments.

10.3.5 *Hybrid sensors*

A hybrid optical sensor comprises a conventional sensing element together with an electrical–optical converter for transmitting the sensor data along a fibre. The sensor electronics is of low power consumption and is powered either from local batteries or from an optical fibre providing electrical power via a photodiode. Figure 10.14, (a) and (b), shows two possible systems.

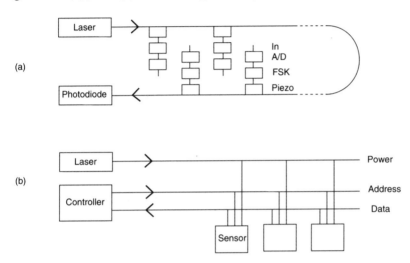

Figure 10.14 *Hybrid sensor systems.*

In figure 10.14(a), the sensors have local batteries and the data (in the form of bursts of sinewaves of different frequency representing '1' and '0') are converted into optical signals by a piezo. In figure 10.14(b), the sensor excitation is provided optically, and the arrangement is similar to that in a computer.

The sensors in figure 10.14, (a) and (b), are conventional devices (non-resonant) but resonant sensors have recently been used in a similar arrangement, shown in figure 10.15. This is a sort of 'ring main' system, consisting ideally of a single optical fibre carrying both power and signals. The sensors would be resonant devices, each with a different basic frequency. The sensors may operate continuously or may be addressed by modulating the optical power at suitable frequencies to excite just one sensor.

It has proved difficult to operate multi-sensor systems but several optically excited sensors have been reported, including a temperature sensor (Spoonser *et al.*, 1987) and a vibrating wire force sensor (Jones and Philp, 1983), shown in figure 10.16. Such systems usually use two fibres, one for driving and one for receiving, and employ resonant sensors maintained in vibration by a small amount of optical power.

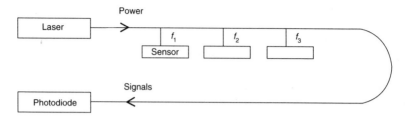

Figure 10.15 *'Ring main' system for resonant sensors.*

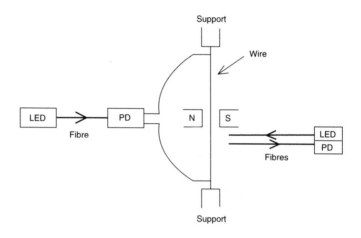

Figure 10.16 *Optically-powered vibrating wire force sensor.*

The inherent power of fibre-optic transducers has not yet been fully exploited. They are not ideally suited to interfacing to most microprocessors, though opto-electric conversion is fast and efficient. They do not at present utilise the enormous information capacity of optical fibres, nor do they make use of the ability of optical systems to perform 'instantaneous' Fourier transforms. However, it is clear that these features will be developed in the future, when multiplexed transducers, parallel data transmission and 'instantaneous' processing will offer remarkable possibilities. If 'optical computers' are developed, they would of course be ideally suitable. A comprehensive treatment of optical fibre sensors and systems has been given by Culshaw and Dakin (1989).

10.4 Pyrometry

The Greek word *pyro* means *furnace*, and pyrometry is the remote measurement of the temperature of a body by measuring the radiation emitted. Its uses include the measurement of furnace and flame temperatures, detection of fires or overheated components, intruder detection, missile guidance and the location of crop diseases (used from aircraft) and ore concentrations (used from spacecraft). Radiation from a hot body is known as 'black body radiation' and its properties are summarised below.

10.4.1 Black body radiation

Kirchhoff's law

If radiation W falls on a body, as in figure 10.17, a fraction α_a is absorbed and a fraction α_r ($= 1 - \alpha_a$) reflected. The body will tend to increase in temperature, emitting radiation ϵW, where ϵ is a constant known as the emissivity, until thermal equilibrium is reached. In this condition $\alpha_a = \epsilon$, and it can be shown that this is true at all frequencies ν. For a *black body* for which $\alpha_r = 0$, $\alpha_a = \epsilon = 1$ and all radiation is absorbed and re-emitted. The emissivity ϵ is defined as the ratio of absorbed to incident power and is numerically the same as the absorbivity α_a.

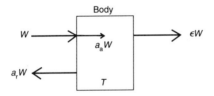

Figure 10.17 *Kirchhoff's law.*

Stefan's law

This states that the total radiant power W emitted per unit area by a body at temperature T Kelvin is given by $W = \epsilon\sigma T^4$ W/m^2, where ϵ is the emissivity and σ is the Stefan–Boltzmann constant, of value 5.6×10^{-8} W/(m^2 K^4).

Planck's law

Planck's law gives the distribution of radiant power against wavelength, shown in figure 10.18. Attempts to predict this using classical physics produced the so-called *ultra-violet catastrophe,* with power increasing indefinitely at short wavelengths, and this led ultimately to the development of quantum theory.

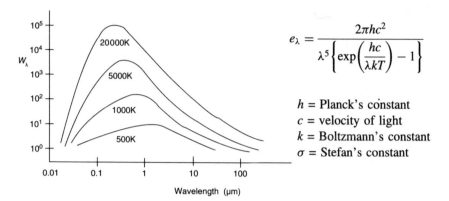

$$e_\lambda = \frac{2\pi hc^2}{\lambda^5\left\{\exp\left(\dfrac{hc}{\lambda kT}\right) - 1\right\}}$$

h = Planck's constant
c = velocity of light
k = Boltzmann's constant
σ = Stefan's constant

Figure 10.18 *Plank's law.*

The curve has a long 'tail' at long wavelengths, and retains its shape with increasing temperature, though the peak moves to shorter wavelengths. About a quarter of the total power lies on the short-wavelength side of the peak. The expression for Planck's law is interesting in that it includes four different universal constants. The shape is governed by the ratio of photon energy $h\nu$ to equipartition energy kT. Integration over all wavelengths yields Stefan's law, so σ is not an independent universal constant, unlike h, c and k. The integration is not recommended to persons of sound mind.

Lambert's law

This gives the distribution in the direction of the radiation from a hot body. In figure 10.19, the radiation δW at an angle θ to the normal to the hot body of area δa in a solid angle $\delta\omega$ is given by

$$\delta W = \frac{W \cos\theta}{\pi}\delta a\,\delta\omega$$

where W is the emitted radiation, given by either Stefan's law or Planck's law as appropriate. This expression is very useful in pyrometry in evaluating the radiation received by a detector at a known distance from a hot body.

Wien's law

By differentiating Planck's law it can be shown that the wavelength λ_m corresponding to peak emission is given by $\lambda_m \times T = 2.9 \times 10^{-3}$ m K. λ_m is therefore approximately $2900/T$ μm so radiation at room temperature of 290K has

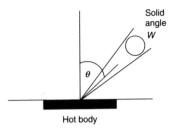

Figure 10.19 *Lambert's law.*

a peak emission at 10 μm, and that from the sun (6000K) at about 0.5 μm, unsurprisingly in the middle of the visible spectrum.

10.4.2 Types of pyrometer

(a) Simple total radiation system

If a suitable detector is placed a known distance d from a hot body of known area A and temperature T, the radiation received δW can be estimated easily using Lambert's law. In figure 10.20, the solid angle δW subtended by the detector is $\delta W = a/d^2$ and $\cos\theta = 1$ so

$$\delta W = \frac{\epsilon\sigma T^4}{\pi} \times A\frac{a}{d^2}$$

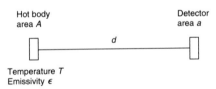

Figure 10.20 *Simple total radiation pyrometer.*

The arrangement can be used for intruder detection, using a pyroelectric detector, or for the detection of overheated electronic components using a lead sulphide cell or a pyroelectric device with a chopper to provide an a.c. signal. However, it is unsuitable for accurate temperature measurement owing to the very wide spectral range of the radiation and the losses occurring between hot body and detector.

(b) Basic pyrometer measurement

Figure 10.21 shows a detector of area A_d receiving radiation from a hot body (target) of area A_t and restricted by an aperture of area A_a containing a lens.

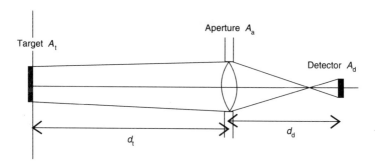

Figure 10.21 *Basic pyrometer system.*

From Lambert's law (see above), the power falling on the lens is

$$\delta W = \frac{W \cos\theta}{\pi} \delta a \delta W = \frac{W}{\pi} A_t \frac{A_a}{d_t^2}$$

If the radiation is focused over the detector area, $\sqrt{(A_t/d_t)} = \sqrt{(A_d/d_d)}$ and

$$\delta W = \frac{W}{\pi} \times \frac{A_d A_a}{d_d^2}$$

This is independent of target distance d_t, so if the target is moved and the radiation focused over the detector area the same radiation is received (though the target area changes, of course). This enables accurate remote measurement of temperature without having to know the target distance, and is the basis of all practical instruments. There are two main types: total radiation pyrometers, which use Stefan's law, and monochromatic pyrometers, using Planck's law. The former suffer from the disadvantages of the simple system described above and are little used for accurate temperature measurement. A monochromatic type is shown in figure 10.22.

A narrow band of radiation (for example 0.01 μm at 0.6 μm) is selected by a filter and the amount of radiation collected by the lens can be calculated using Lambert's law with W given by Planck's law. In the original instrument the radiation was focused on to a heated filament which 'disappeared' in the field of view when its brightness was the same as the radiation received. In modern instruments this process is done automatically by alternately reflecting the radiation received and that from a reference lamp on to a suitable detector, such

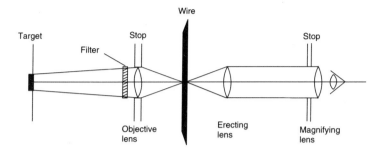

Figure 10.22 *Monochromatic pyrometer.*

as a *p–n* photodiode (in the visible region) or a lead sulphide cell at lower temperatures.

10.5 Ultrasonic measurement systems

The word *ultrasonics* refers to frequencies higher than the response of the ear (about 18 kHz) and typically uses frequencies in the range 40 kHz to 1 MHz. Piezoelectric transducers are almost always used both for the production and reception of the ultrasonic wave. This may be either a continuous sinewave, as in some flow measurement systems, or a short burst of sinewaves as in most distance measurement systems. A typical arrangement is shown in figure 10.23. The inverse piezoelectric effect ($x = dV$) is used at the transmitter and the normal effect ($q = dF$) at the receiver.

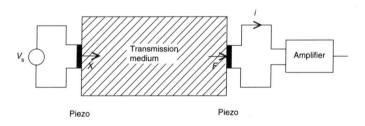

Figure 10.23 *Basic ultrasonic system and equivalent circuit.*

10.5.1 Propagation of ultrasonic waves

Ultrasonic waves have velocities ranging from about 6 km/s for steel down to about 0.3 km/s for air. The attenuation of a plane wave depends very strongly on the medium, but can be described by a power attenuation coefficient α such that the power P_x at distance x is given by $P_0\exp(-\alpha x)$. The behaviour at boundaries between media is determined by the characteristic impedance R_a, given by the product of density ρ and velocity c. For near normal incidence of a wave at a boundary between two media of impedance R_1 and R_2, the reflection and transmission coefficients, α_r and α_t respectively, are given by

$$\alpha_r = \frac{(R_2 - R_1)^2}{(R_2 + R_1)^2}, \qquad \alpha_t = \frac{4R_1 R_2}{(R_2 + R_1)^2}$$

Some values of R_a for typical materials are given in table 10.3 and values of α_r and α_t for common boundaries in table 10.4.

Table 10.3

Material	Velocity (m/s)	Density (kg/m³)	Impedance (kg/m² s)
Quartz	5.6×10^3	2.65×10^3	1.5×10^7
BaTiO₃	4.4×10^3	5.7×10^3	2.5×10^7
Steel	6.0×10^3	7.8×10^3	4.7×10^7
Water	1.5×10^3	1.0×10^3	0.15×10^7
Bone	5.3×10^3	1.5×10^3	0.8×10^7
Tissue (biological)	1.5×10^3	1.0×10^3	0.15×10^7
Polystyrene	2.4×10^3	1.1×10^3	0.26×10^7
Air	0.34×10^3	1.3	440

Table 10.4

	Steel		Quartz		Bone		Water		Polystyrene		Air	
	α_r	α_t	α_r	α_t	α_r	α_t	α_r	α_t	α_r	α_t	α_r	α_t
Steel	0	1	0.27	0.73	0.5	0.5	0.88	0.12	0.8	0.2	0.99	4×10^{-5}
Quartz			0	1	0.09	0.9	0.67	0.33	0.5	0.5	0.99	1×10^{-4}
Bone					0	1	0.47	0.53	0.24	0.76	0.99	2×10^{-4}
Water							0	1	0.07	0.93	0.99	1×10^{-3}
Polystyrene									0	1	0.99	7×10^{-4}
Air											0	1

The fraction of ultrasonic energy received in a system can be evaluated easily using these values. For example, consider the two paths shown in figure 10.24, for measuring the depth of water in a steel pipe.

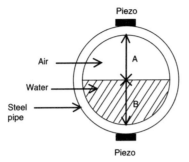

Figure 10.24 *Transmission of ultrasonic waves in pipe.*

Ignoring attenuation, the fractions of power received to power transmitted are:

Path A: $(QS)(SA)(AW)_r(AS)(SQ) = (0.73)(4 \times 10^{-5})(0.99)(4 \times 10^{-5})(0.73)$
$$= 8.4 \times 10^{-10}$$

Path B: $(QS)(SW)(WA)_r(WS)(SQ) = (0.73)(0.12)(0.99)(0.12)(0.73)$
$$= 7.6 \times 10^{-3}$$

The efficiency of power transfer into a medium of low acoustic impedance can be improved by the use of a matching layer, and it can be shown that the best matching impedance R is given by the geometric mean of the two materials, that is, $R^2 = R_1 R_2$.

10.5.2 Applications of ultrasonics

(a) Pulse reflection systems

Pulse reflection systems are radar-like systems in which a short burst of sinewaves is emitted and the reflected pulses analysed. Ideally the burst would be very sharply defined, so it can be considered as a rectangular pulse, but in practice the Q of the system limits the rise time. Figure 10.25(a) shows a typical system and figure 10.25(b) the reflected pulses.

The amplitude of the first reflected pulse can be calculated as above, and further reflections occur of decreasing amplitude. If the ultrasonic frequency is f then we have the requirements $T_t \gg 1/f$, $T_r \gg T_t$, and $T_0 \gg T_r$ for accurate and unambiguous measurement of the reflected pulse time T_r. The accuracy is of the order of half the wavelength of the ultrasonic wave, and is about ±5 mm for $f = 40$ kHz, decreasing to about ±0.2 mm at 1 MHz.

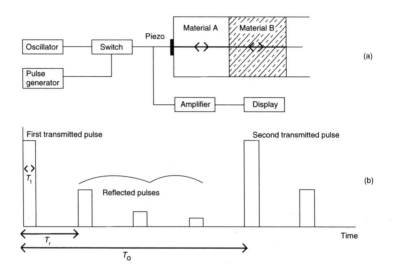

Figure 10.25 *Ultrasonic pulse reflection system.*

Pulse reflection methods have many applications, including distance measurement (in air or water), thickness measurement, detection of flaws in metals, imaging of body tissues and flow measurement by transit-time methods (described in chapter 8).

A number of piezoelectric ultrasonic transducers are available for use in distance measuring systems and are available singly or in pairs. Single types are used both for transmitting and receiving and offer simplicity of mechanical construction, and increased accuracy at short range. Pairs can often offer increased range as the transmitter can be optimised for high power handling and the receiver for high responsivity. The disadvantage of pairs is that they cannot occupy the same space and hence give erroneously high range values at very short ranges. Systems using single transducers also have problems at very short ranges, however, as the receiver cannot begin to receive until the transmitted pulse has finished. There is thus a minimum range below which the range cannot be determined. Neither of these is a problem if the systems are not required to work at very small ranges (typically < 20 mm).

A major problem with ultrasonic systems, when compared to radar systems, is that it is the speed of *sound* in the medium that dictates the echo time rather than the speed of *light*. The speed of sound in air, for example, depends on the air temperature and other parameters. It is possible to make the frequency of the counter that measures the echo time temperature-dependent, so as to compensate for the effects of temperature change, but it is more difficult to compensate for other effects. Another cause of inaccuracy is the non-rectangular shape of the received pulse. For a given threshold level, pulses of small amplitude are detected slightly later than large ones. As distant objects give smaller echoes than close ones, the threshold can be made to reduce with time during each pulse cycle to compensate for this. The effects of differing target sizes still give problems, however, as they cannot be predicted.

Transducers of small diameter, and hence wide beamwidth, are often used in collision-avoidance systems where precise knowledge of the direction of the detected object is not important. Transducers with larger diameters, and hence narrow beamwidth, are more commonly used in mapping systems or as ultrasonic tape measures. Because of the inaccuracies, mentioned previously, these systems are referred to as distance estimators rather than distance measurers. Phased arrays of transducers may be used to scan electronically a narrow ultrasonic beam and hence produce a two-dimensional ultrasonic image. These techniques are used in medical ultrasound scanners and in some robot vision systems.

(b) Continuous wave systems

The two most important applications are the ultrasonic Doppler flowmeter and ultrasonic correlation methods for flow measurement.

Figure 10.26 shows an ultrasonic Doppler flowmeter in which continuous sinewaves are injected into a liquid and the Doppler frequency shift measured. This is given by $\delta f = (2f/c)\cos\theta\, v$, as for the laser flowmeter, but δf is much lower of course, being typically about 20 kHz for a flow rate of 10 m/s in water with $f = 1$ MHz and $\theta = 30°$. δf is measured by mixing the transmitted and reflected waves, when the amplitude appears modulated at frequency $\delta f/2$.

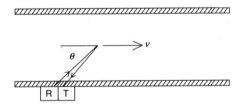

Figure 10.26 *Ultrasonic Doppler flowmeter.*

The method depends on there being suitable scattering centres in the fluid, but has the advantage that it is essentially non-invasive and 'clip-on', being simply attached to the outside of the pipe.

Figure 10.27 shows an ultrasonic correlation system. Again, the method requires the presence of bubbles or eddies in the flow, but is completely non-invasive.

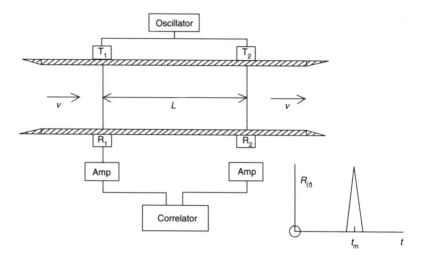

Figure 10.27   *Ultrasonic correlation flowmeter.*

The flow velocity $v = L/\tau_m$, as for the optical device in chapter 8.

10.6   Exercises

10.6.1. (a) Discuss the properties of optical fibres which may be used for sensing purposes, and the advantages and disadvantages of such systems.

(b) Explain what is meant by the optical processing of data and the extent to which such processing is now possible.

10.6.2. (a) Discuss the advantages and disadvantages of optical fibre sensors for measuring small displacements up to 1 mm, compared with capacitive displacement transducers.

(b) A reflective optical fibre sensor using two fibres is to be used to measure the displacement of a cantilever beam in a force measurement system. The graph of reflected optical power versus displacement is shown in figure 10.28. Estimate the responsivity of the measurement of displacement, assuming a photodiode of responsivity 0.1 A/W is used to detect the reflected light over the linear section AB of figure 10.28. Estimate the detectivity in a bandwidth of 100 Hz in mid-range (point C of figure 10.28), assuming it to be dominated by the shot noise in the diode.

10.6.3. Discuss the advantages of resonator sensor systems and the fields of measurement in which the idea can be applied. Is a system that produces information in the form of a frequency proportional to the desired parameter necessarily superior to one producing a proportional voltage?

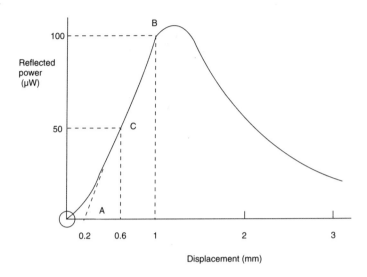

Figure 10.28

10.6.4. (a) Discuss the properties of silicon that make it suitable for the construction of large numbers of very small circuits, and its suitability as a material for the construction of sensing elements.

(b) It has been suggested that the next step in evolution may be the development of super-intelligent self-replicating 'creatures' based on silicon, and that the development of organic creatures culminating in man was simply a necessary step in the process. Discuss the validity of the premises on which the theory is based.

10.6.5. A small probe is required for detecting and locating overheated components in electronic circuits, by measuring the infra-red radiation emitted. The photodetector used should have an effective diameter of about 2 mm and be held about 3 cm from the component being checked. Discuss the design of the system, making reasonable estimates of the parameters involved, and estimate the responsivity of your proposed system.

10.6.6. (a) The depth of water flowing in a steel pipe of diameter 20 cm and thickness 5 mm is to be measured by an ultrasonic pulse reflection method as shown in figure 10.29, using a quartz transmitter/receiver of natural frequency 1 MHz.

(i) Calculate the ratio of received-to-transmitted power and the 'round trip' time using the data below, assuming the water has a depth of 10 cm and attenuation coefficient 5.0 m^{-1}. State any assumptions made.

(ii) Suggest suitable values for the pulse width and for the pulse repetition rate, and estimate the accuracy of measurement.

(b) The velocity of the water in figure 10.29 is to be measured by an ultrasonic Doppler method, using the same transducer position and injecting the beam at an angle of 30 degrees to the flow.

(i) Calculate the frequency shift of the reflected beam for a water velocity of 1 m s^{-1}.

(ii) Discuss the relative merits of velocity measurements by ultrasonic Doppler and transit time methods for this application.

	quartz	*steel*	*water*	*air*
Velocity of sound (m/s)	5.5×10^3	6.1×10^3	1.5×10^3	0.34×10^3
Density (kg/m^3)	2.5×10^3	8.0×10^3	1.0×10^3	1.3

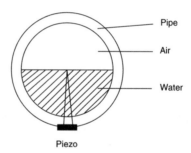

Piezo

Figure 10.29

11 Digital Transducers and Interfacing

As discussed in chapter 1 of this book, the quantities we wish to measure are usually inherently analogue in nature, and most so-called digital transducers simply employ well-known analogue effects in such a way as to produce a digital output. Various digital devices have therefore been discussed in the chapters on length, temperature or light measurement. In this chapter we will first discuss the concept of digital measurement, then make a classification of digital transducers and finally discuss the interfacing of transducers in computer systems.

11.1 Digital measurements

The digital measurement of a quantity means that its value is represented by a certain number of digits, usually decimal, as provided by a measurement of voltage with a 4-digit digital volt meter (DVM) or of frequency by a 6-digit electronic counter. The *resolution* of such a measurement is one part in the displayed number in the time taken for the measurement, for example 1 in 10^6 in 1 s. In practice, the useful resolution is such that the noise level in the system is about equal to the least significant digit.

A digital measurement usually involves either measuring a frequency or counting the number of pulses in a given time, and the basic process is illustrated in figure 11.1. The standard oscillator is an accurate crystal-controlled device (typically 1 MHz or 10 MHz) providing the reference for the measurement, and precise time intervals are generated by dividing this frequency by a set of decade dividers.

In the frequency mode of operation, the gate is held open for a time determined by the dividing chain and the number of pulses passing through the gate into the counting chain is a measure of the input frequency. For example, if the input frequency is 1 kHz and the gate time 1 s (generated by dividing a standard oscillator of 1 MHz by six decade dividers) the counted number would be 1000 and the resolution 1 in 10^3 in 1 s. In frequency mode, the resolution is always equivalent to 1 in f in 1 s, where f is frequency, so it will be low for low values of f unless a long gate time is used.

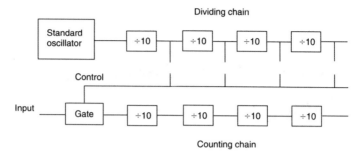

Figure 11.1 *Frequency mode measurement.*

In the period mode of operation, the input frequency is fed to the dividing chain and the standard oscillator to the gate, as in figure 11.2. The gate is held open for the time taken for a preset number of input cycles, so for an input frequency of 1 kHz and a 1 s gate time the resolution is 1 in 10^6 in 1 s (for a 1 MHz standard oscillator) which is much higher than the 1 in 10^3 in 1 s in frequency mode. A high resolution can always be obtained in period mode, though with the disadvantages of proportionality to period rather than frequency, and a much greater susceptibility to noise. A time interval can be measured in exactly the same way, counting the number of standard oscillator cycles in the time t_g for which the gate is open, with resolution 1 in $10^6 \times t_g$.

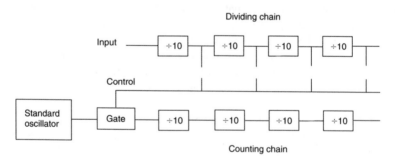

Figure 11.2 *Period mode measurement.*

It is always possible to obtain a high resolution (say, 1 in 10^6 or better) when measuring a frequency or time interval, by using the appropriate measurement mode, and modern electronic counters often select the best mode automatically.

11.2 Digital transducers

A digital measurement could, of course, be obtained by feeding an analogue output to an analogue-to-digital converter (ADC). Cheap 12-bit ADCs are readily available though 16-bit devices are relatively expensive. However, higher resolutions can usually be obtained by a device that produces a frequency or time-interval output, utilising the capabilities of electronic counters discussed above. A frequency or time-interval itself is not strictly digital, of course, but most of the work of conversion has already been done, and devices producing such outputs are usually known as digital transducers. There are three main classes of digital transducers: frequency output devices, time-interval output devices and inherently digital devices.

11.2.1 Frequency output devices

(a) Analogue devices

The simplest devices are those in which an analogue transducer is operated in such a way as to produce an output frequency. For example, a variable-area capacitive displacement transducer can be arranged so that the capacitances between the fixed and moving plates control two oscillators, as shown in figure 11.3. A mixer then extracts the difference frequency $(f_1 - f_2)$. In practice, an offset is applied so that $(f_1 - f_2)$ remains positive and to avoid problems of frequency 'locking'. $(f_1 - f_2)$ can be measured with high resolution in period mode.

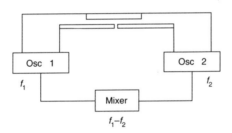

Figure 11.3 *Variable-area capacitive sensor with frequency output.*

Similarly, temperature can be measured using a digital quartz thermometer, in which a specially cut quartz crystal in an oscillator circuit is exposed to the required temperature, and produces a frequency change of about 1 Hz per °C centred on 5 MHz. A high resolution can be obtained by mixing with another oscillator at, say, 5 001 000 Hz and measuring the 1 kHz difference frequency in period mode.

(b) Resonator sensors

Resonator sensors have been described in detail in chapter 10. They involve vibrating wires or beams arranged such that their natural frequency is proportional to the desired measurand such as pressure or force. The frequency is measured in frequency or period mode, as required to provide a suitable resolution.

11.2.2 Time-interval output devices

These also employ analogue transducers, but arranged to produce a time interval. The technique is similar to that used with some types of ADC in which the time taken for an integrator to reach a preset level is measured. Figure 11.4(a) shows a resistive potentiometer for displacement measurement used in this way.

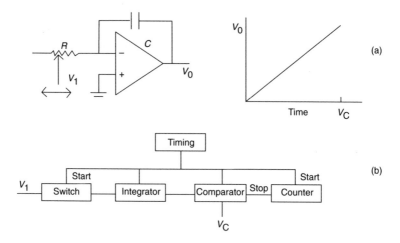

Figure 11.4 *Resistive sensor producing time-interval output (a), and control system (b).*

A reference voltage V_1 is applied to the slider, and the output $V_0 = 1/(RC) \int V_1 \, dt$ rises linearly until it reaches a preset level detected by a comparator. The complete arrangement is shown in figure 11.4(b). The same principle can be used with many types of transducer such as strain gauges, photoconductive cells, thermistors, etc.

11.2.3 Inherently digital devices

A small number of devices or techniques are best classified under this heading, though they do, of course, employ essentially analogue techniques.

(a) Encoders

Encoders for angle measurement have already been described in chapter 5. Absolute types consist of discs with coded tracks with the angular position read optically, and provide a low resolution equal to 1 in 2^n, where n is the number of tracks. Incremental types simply produce a number of pulses proportional to the angle turned through, with corresponding resolution.

(b) Grating measurement systems

Moiré grating displacement systems were also discussed in chapter 5, and consist of a fixed and a moving grating, the number of intensity cycles over the required displacement being counted. The displacement resolution is basically the reciprocal of the line spacing (for example, 10^{-4} m with a grating of 10 lines per mm), though interpolation by a factor of 10 or more can often improve this.

(c) Interferometric systems

An interferometric system is similar to a grating system in that the number of cycles of intensity corresponding to the distance traversed is measured. However, the displacement resolution is determined by the wavelength of light rather than the grating rulings. Figure 11.5 shows a Michelson interferometer in which interference occurs between beams of light traversing the two paths shown. A Helium–Neon gas laser or laser diode provides coherent light and the interference fringes are detected with a solar cell and fed to a counter. Special mirrors (cube corner reflectors) are used to simplify alignment and a displacement resolution of 0.3 μm (corresponding to half the wavelength of light at 0.6 μm) can be obtained over a distance of many metres.

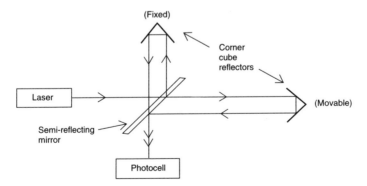

Figure 11.5 *Interferometric displacement measuring system.*

Other types of interferometer are in use, though the basic principles and resolution are similar.

11.3   Interfacing

11.3.1   Introduction

Nowadays, most measurement systems are connected to a computer at some stage. This is often for data collection or data processing purposes, but in some cases may be concerned with the control of the measurement process itself. The data may be digital or analogue, and may be available in serial or parallel form. For example, an incremental encoder produces a sequence of digital data though this is usually fed to a counter, producing parallel data in the form of several decimal digits, whereas a ten-track absolute encoder directly produces ten binary digits in parallel. An analogue transducer, on the other hand, produces an analogue voltage or current, which must be digitised before it can be fed to the computer.

The computer may typically be a PC, with keyboard and display, or may be a single chip device wired on the same circuit board as the measurement system. A simple block diagram of a general computer system is shown in figure 11.6. It comprises a central processing unit (CPU), memory and an input–output (I/O) unit, connected by a collection of wires known as a 'bus'. The memory contains both instructions and data, and the basic operation is that an instruction is 'fetched' from memory and decoded and executed by the CPU. The whole process is synchronised by a 'clock', each individual operation requiring one or more clock cycles. The process continues until the sequence of instructions is completed. The clock cycles on modern computers are extremely short, so that data are output or must be input in a very short time, and standardised chips are used to accomplish the process.

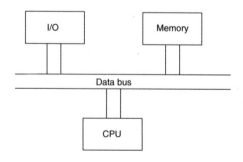

Figure 11.6   *General computer system.*

11.3.2 Input and output ports

We will consider first the schematic interface circuit shown in figure 11.7, which provides two 8-bit 'ports', one for input and one for output.

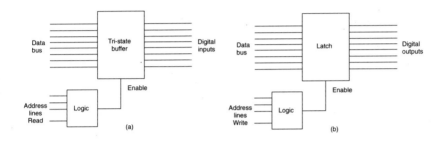

Figure 11.7 *Simple input port (a) and output port (b).*

For the input port it must be arranged that the data from the external device are only connected to the data bus when required, so that the data lines are not permanently held in one condition, and this is achieved by a 'tri-state' buffer. When enabled it can pass an input of 0 or 1 to its output, but when in the tri-state condition it presents a very high resistance to the data bus and does not affect its operation. The enable input must be activated only when the appropriate memory address is referenced, and the address lines are decoded to produce a suitable control signal.

The data on the bus are stable for only a very short time and must therefore be sent to the output port during this time and held there until required. This is accomplished by a 'latch' which accepts data only when enabled and retains it irrespective of changes to its input until enabled again.

Customised chips providing input and output ports, and usually other facilities such as counters and timers, are available from several manufacturers. Typical examples are the 6522 Versatile Interface Adapter (VIA) and the 8255 Parallel Peripheral Interface (PIA). The pin connections for the 6522 are shown in figure 11.8. It provides two 8-bit ports in which each bit may be individually programmed as either input or output by setting the appropriate bits in the data direction registers. It also includes two programmable 16-bit timers and an 8-bit shift register for serial data transfer, and facilities for 'interrupts', whereby the operation of the main program is suspended while the processor carries out a special routine (such as reading in some data, or updating a time-of-day register).

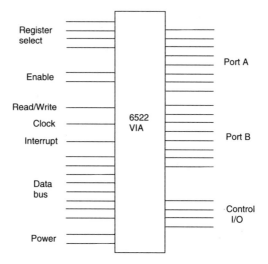

Figure 11.8 *Versatile interface adapter (VIA).*

11.3.3 *General-purpose interface boards*

Many manufacturers produce interfacing boards that plug directly into a PC and provide a range of facilities, usually including input and output ports, analogue-to-digital conversion (ADC), digital-to-analogue conversion (DAC) and timers. They often contain a VIA in addition to ADC and DAC chips. A schematic of a typical board is shown in figure 11.9.

The address decoder chip is fed from the address lines via logic gates so as to derive the required enabling signals, and the bus transceiver determines the direction of data flow to or from the output or input port. The 'base' address of the board is set by the address decoding logic and the addresses for the digital inputs or outputs, ADC and DAC, are then related to this base address. In this schematic example the addresses are as follows:

$$
\begin{array}{ll}
\text{Base address :} & \text{\$C0C0 (decimal 49344)} \\
\text{ADC} & \text{base} + 0 \\
\text{DAC} & \text{base} + 1 \\
\text{DIG In} & \text{base} + 2 \\
\text{DIG Out} & \text{base} + 3 \\
\end{array}
$$

Digital inputs and outputs   To read in a digital input, for example from a binary switch or a set of such switches, it is necessary only to perform a 'read' operation to the relevant address. The instruction in BASIC

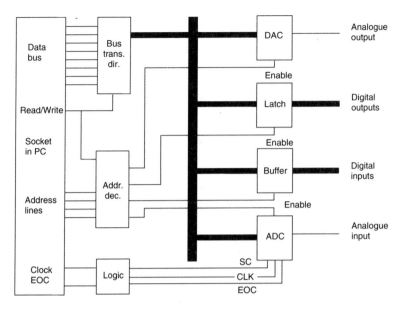

Figure 11.9 *Schematic interface board.*

$$10 \ VALUE \ = \ INP \ (base \ + \ 2)$$

reads all eight bits of the port together, and individual bits must then be extracted if required. Similarly, the instruction

$$20 \ OUT \ (base \ + \ 3), \ VALUE$$

sends digital outputs corresponding to VALUE to the output port.

Digital-to-analogue conversion Digital-to-analogue converters usually employ a 'ladder' arrangement in which the individual digits are weighted by suitable resistors to produce the analogue output. The conversion is almost instantaneous and is performed by an instruction of the same form as for a digital output, for example

$$30 \ OUT \ (base \ + \ 1), \ VALUE$$

Analogue-to-digital conversion There are many types of ADC devices, but the most widely used type is the successive approximation ADC. This contains a DAC and a register, with all bits initially set to '1'. Starting with the most significant bit, the output of the DAC is compared with the input and the bit retained or set to zero as required. Successive bits are then added and the process continued until all the bits have been tried. The particular advantage is that each comparison takes only one clock cycle, so the process is very fast. The process is initiated by a 'write' operation to the relevant address, followed by a 'read' operation. The best method is to monitor the 'end-of-convert' signal and 'read' as soon as it changes state to signal that conversion is complete. However, BASIC is so slow that we need not bother, and the instructions below are all that is needed:

```
40 OUT (base),  1
50 VALUE  =  INP (base)
```

Commercially available boards may cost anywhere between about £100 and £1000 depending on the facilities available. A typical board may provide the following:

16 single-ended analogue inputs with 12-bit resolution

2 analogue outputs with 12-bit resolution

16 digital input channels

16 digital output channels

Programmable interval timer

A range of base addresses is usually selectable by the user by means of 'jumpers'. The ADC is usually a successive approximation type with conversion time say 30 µs and a multiplexer is used to obtain the 16 ADC channels. Eight-bit boards are rarely used now, most boards being 12-bit or 16-bit, though the latter are relatively expensive and considerable precautions in wiring and earthing are necessary to obtain 16-bit resolution in practice. The 12- or 16-bit boards still operate with 8-bit bytes, but contain more registers so that the upper and lower bytes are read out separately. The program section below reads in one channel of the ADC:

```
10 BASE  =  $H200
20 OUT BASE +  10,  CHANNEL 'selects channel'
30 OUT BASE +  11,  1            'starts conversion'
40 HIBYTE  =  INP (BASE + 5)
50 LOBYTE  =  INP (BASE + 4)
60 VALUE  =  (HIBYTE * 256 + LOBYTE)
```

11.3.4   Instrument interfaces: the IEEE-488 bus

The most widely used interface for the connection of instruments is the IEEE-488 or General Purpose Interface Bus (GPIB). It allows up to fifteen instruments to be connected to one another and to a controlling computer. It would be possible to add the interface to a specific transducer, but there would be little point as the transducer data could more easily be read in as described above, and usually would be fed to a counter or digital voltmeter in any case. Devices are classified as 'controllers' (such as a computer), 'talkers' (capable of sending data, such as a computer) or 'listeners' (capable of receiving data, such as a digital voltmeter), and a given device may have all three functions. A typical arrangement is shown in figure 11.10.

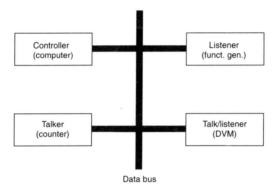

Data bus

Figure 11.10 *Instruments in IEEE-488 system.*

The bus comprises 24 lines, 8 for data, 3 for data transfer, 5 for management and 8 ground lines. The data lines are used for both data and instructions and operate in a bit-parallel, byte-serial manner. The three data control lines (NRFD – not ready for data, NDAC – not data accepted, and DAV – data available) are referred to as a three-line handshake system, and permit efficient transfer of data between devices of widely different data rates. The management lines include: ATN – attention (indicating whether a byte is data or a command), EOI – end or identify, IFC – interface clear, REN – remote enable (allowing a device to be remotely controlled) and SRQ – service request. The latter is pulled down by any device requiring service (generating an error, etc.) and a 'poll' can be carried out to examine the device. A poll can be 'serial', producing a byte of information but only from one device, or 'parallel', checking the SRQ lines of up to 8 devices.

The signal levels, logic, cables, plugs/sockets, etc. are all standardised, special 'piggy-back' plugs being used. Any instruments with the interface fitted can be immediately connected, either in series or star configurations to a controlling computer. Instruments have a Device Address (between 1 and 30) set by five switches on the rear panel. A given instrument is capable of certain functions which are usually stamped on the rear panel. These include Source Handshake

(SH), Acceptor Handshake (AH), Talker (T), Listener (L), Service Request (SR), Remote/Local (RL), Parallel Poll (PP), Device Clear (DC), Device Trigger (DT) and Controller (C). There are different levels of implementation of the various functions and a typical DVM is marked SH1, AH1, T5, L3, SR1, RL1, PP1, DC1, DT1, C0, indicating that most functions are available except that of a controller.

Data and commands are transmitted on the bus as bytes in ASCII and each instrument has its own specific code (of the manufacturer's choice) for its particular operations. For the DVM referred to above, the string 'M1N0R0T0Q0G' sets an a.c. voltage range, numeric display with literals, autorange, sample mode, SRQ on error and trigger. Similarly on a frequency synthesiser, the string 'W1F51.5A01.9D05' produces a sinewave of frequency of 51.5 kHz, amplitude 1.9 V a.c., and d.c. offset 0.5 V. The program line required in BASIC for the first string above (referred to as C$) to the Listener with address L$ is simply:

100 PRINT 'WT'; L$; C$

Similarly, to accept data from a Talker of address T$, placing it in a string S$, requires

200 PRINT 'RD'; T$: INPUT S$

The interface is an excellent solution to the problem of controlling a number of instruments from a central computer, and although developed in the late 1970s it is still widely used. It has the disadvantages of limited speed, of being only 8 bits wide and of the cost associated with both the interface on each instrument and the card required by the controlling PC.

11.3.5   Data communication

In most measurement applications the transducer system is directly interfaced to a microprocessor or PC, as discussed above. However, in some cases such as measurement in hostile environments or in remote locations, such direct interfacing is not possible and the data may have to be communicated to a computer at a convenient location. For distances of more than a few metres the data are transmitted serially, since in principle only two conductors are required compared with eight or more for parallel communication (24 for the IEEE-488 system).

There are two types of serial communication, synchronous and asynchronous. In synchronous systems a clocking waveform is used to ensure correct timing at the transmitter and the receiver, and special synchronisation characters may be included. Asynchronous transmission is more usual in measurement systems, and correct timing is achieved by special start and stop bits, as illustrated in figure 11.11.

Figure 11.11 *Asynchronous data framing.*

The data are arranged in 8-bit bytes, the line being held high when not in use. The start of the byte is indicated by a negative edge, producing the start bit, followed by the 8 data bits and terminated by one or more stop bits. The time for one bit must be accurately maintained to ensure accuracy of transmission, and the reciprocal of the bit time is known as the *baud rate*, a typical value being 9600 baud. The data are usually in ASCII form (7 bits) with an additional parity bit for error detection.

The standard chip for serial data communication is the *Universal Asynchronous Receiver/Transmitter*, somewhat unsurprisingly referred to as the *UART*. It consists essentially of shift registers with suitable control circuitry, as shown in figure 11.12.

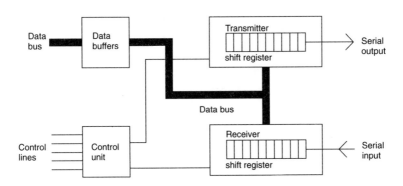

Figure 11.12 *Universal Asynchronous Receiver/Transmitter (UART).*

The parallel data from the transducer are fed to the shift register and, when full, the bits are transmitted serially down the line, another eight parallel bits being then fed in. At the other end of the line another UART converts the serial data stream to parallel, ready for feeding to the computer. In practice, additional lines or signals are required to ensure that the receiver is ready for data, and hardware or software means may be used.

The most widely used standard for serial data communication is the Electronic Industries Association's RS232-C, which specifies the logic levels and mechanical connectors to be used. The standard is rather messy and ambiguous, specifying no less than 25 lines compared to the two required in principle, and is mainly used for serial transmission between two computers, or between a computer and printer or other peripheral device at baud rates up to about 20 000. Other standards, such as RS-422 and RS-423, are used for higher speeds.

11.4 Smart sensors

Although the word 'smart' is somewhat misused, the meaning is reasonably clear. The term refers to transducers in which intelligence is added by means of a digital processor, preferably on the same circuit board. Single chip processors are widely used in domestic and industrial equipment. The Zilog Z8601, has 2 k bytes of ROM, 128 bytes of RAM, 32 I/O lines and can be extended to handle up to 62 k bytes of program/data storage.

A general scheme has been suggested by Brignell, and is shown in figure 11.13.

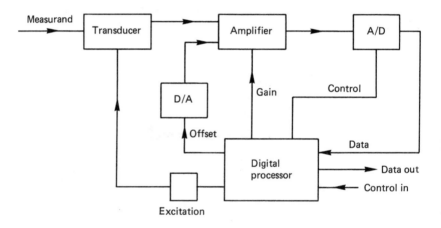

Figure 11.13 *Smart sensor.*

The transducer is under total processor control. The excitation is produced and modified if necessary, depending on the range required. The offset (of the transducer or amplifier) is controlled, as is the gain, and the ADC is similarly adjusted. The processor contains full details of the transducer characteristics in ROM, enabling the correct excitation and gain etc. to be maintained under all conditions. It also contains details of responsivity, including non-linear behaviour, so that the final output is always a correct transduction of the measurand. Indeed, it is already becoming accepted that non-linearity and offsets in transducers are no longer serious shortcomings.

In particular, it is apparent that the actual type of transducer in figure 11.13 is of little importance. Its physical characteristics need to be known and understood, and the basic transducer responsivity and detectivity still determine the overall detectivity. However, whether or not the transducer is actually a so-called digital type is largely irrelevant. The processing would be slightly different and the ADC would be replaced by a scaling device, but all the processing and control would be invisible to the user.

As we pointed out earlier, nearly all transducers are really analogue anyway, but sometimes it is convenient to operate them in a digital manner. We are analogue animals in an analogue world, which is where we started in chapter 1, except that the writer's two word processing fingers are a bit shorter now.

11.5 Exercises

11.5.1. (a) Explain the terms *accuracy, resolution, range* and *measurement time* applied to digital measurements and explain how they are related.

    (b) The frequency of the a.c. mains (50 Hz) is to be monitored with an accuracy of 1 part in 10^4 and a resolution of 1 part in 10^5. Variations in frequency occurring in measurement times of not more than 0.1 s are to be displayed on a chart recorder with full-scale deflection corresponding to 0.5 Hz. Describe how the measurement could be performed.

11.5.2. (a) Discuss the various classes of digital transducers, giving examples of each class. Explain why it is often preferable to use an arrangement in which the digitisation is inherent, rather than to add an ADC to an existing analogue transducer.

    (b) A digital output is required for the following measurements. In each case discuss the possible methods and recommend the most suitable:

    (i) the linear displacement of a shaft moving over a range of ±1 cm. A resolution of ±1 μm in 1 s is required

    (ii) the temperature of a heated copper cylinder over a range from 50°C to 100°C with resolution 0.1°C

    (iii) the light level in a factory over a range of total radiation of 10 to 100 W/m^2 with resolution 1 per cent.

11.5.3. (a) Discuss the advantages of digital transducers over analogue transducers.

    (b) Describe how the following measurements could be made, using a digital transducer in each case (*not* an analogue transducer followed by an ADC):

    (i) the angular position of a shaft moving over a range of ±20° with maximum angular velocity 0.1 rad/s with resolution 10 seconds of arc

    (ii) measurement of the angular velocity of a shaft rotating at about 10 rev/s with resolution 1 part in 10^4 and measurement time 0.1 s.

11.5.4. Figure 11.14 is a schematic diagram of a general-purpose interface for a microcomputer, providing an ADC, DAC, digital outputs and digital inputs. Discuss the operation of each device, explaining its interfacing requirements. Explain how to perform the following operations, either in words, by flow diagrams or in computer code:

(i) output the number 127 to the DAC

(ii) read a byte from the ADC

(iii) output the decimal number 17 to the digital outputs

(iv) determine whether the most significant bit of the digital input is 1 or 0.

Note. The DEVS line is active for a hexadecimal address in the range C000 to C0FF.

The decoder is a 3 to 8 device with lines selected by S_0, S_1, S_2.

The ADC is a successive approximation type. Its clock is provided by dividing microprocessor clock by 4.

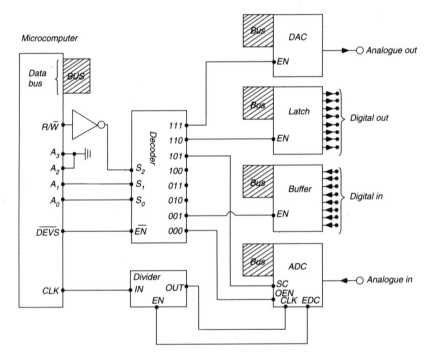

Figure 11.14

11.5.5. Explain the principal features of the IEEE 488 general-purpose instrument bus, and discuss its advantages and disadvantages.

It is required to measure the characteristics of a large number of similar networks. Each network resonates at a frequency between 0.5 Hz and

1.5 Hz and the frequency response is required over the range 0.1 Hz to 10 Hz with an input signal of 1 V and resonance frequency accuracy ±0.1 Hz. An amplitude response is required (at the resonance frequency) over the range 0.1–10 V. A frequency synthesiser, digital voltmeter and printer are to be used, controlled via the IEEE 488 bus by a microcomputer.

Discuss the features of these instruments and give a flow diagram of a suitable program to perform the measurements.

11.5.6. Explain what is meant by a *smart* instrument, discussing the advantages over conventional instruments.

Figure 11.15 shows a proposed arrangement for monitoring the thickness of polythene sheet, of nominal thickness 0.1 mm, with accuracy ±1 per cent. The sheet passes between a smooth fixed base A and a rotating roller B. Roller B is attached to a pivot, and moves up and down in response to changes in sheet thickness, the movement being measured by a displacement transducer C. The angular position of roller B is to be monitored in order to allow for errors in thickness measurement caused by its non-circularity (say 0.01 mm).

Discuss suitable transducers for C and the roller B with regard to the accuracy required, and explain how the data could be used with an on-line microprocessor to display sheet thickness. Comment on the feasibility of the system and explain how it could be calibrated.

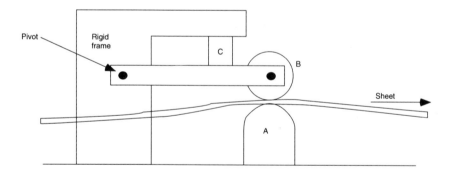

Figure 11.15

Solutions to Exercises

Solutions to chapter 1 exercises

1.7.1. *Responsivity* is response per unit input.
Detectivity is output signal-to-noise ratio per unit input.
Range is the range of input levels over which the responsivity satisfies some stated requirement.
(a) An LVDT (linear variable differential transformer).
Responsivity typically 100 V/m per volt excitation.
Detectivity better than 10^6/m (< 1 μm change detectable).
Range ±1 cm (non-linearity < 1 per cent over this range).
(b) A resistance thermometer (Platinum type).
Responsivity typically 0.5 Ω per 100 Ω per °C.
Detectivity typically 1000/°C (limited by noise).
Range −200 to +850°C (1 per cent non-linearity).
(c) A photovoltaic cell (silicon solar cell).
Responsivity typically 0.1 A/W (at 1 μm).
Detectivity 10^{12}/W.
Range 0 to 1 W (upper limit set by heating effects and no precise value can be given).

1.7.2. (a) (i) Modulator
(ii) Self-generator
(iii) Modulator
(iv) Self-generator (thermal–mechanical).
(b) (i) Thermistor or resistance thermometer
(ii) Piezoelectric crystal or tachometer
(iii) Diaphragm-type barometer
(iv) Silicon solar cell
(v) LVDT or capacitive displacement transducer.

Solutions to chapter 2 exercises

2.6.1. Force–flow: force–voltage
velocity–current
mass–inductance
compliance–capacitance

	mechanical resistance–resistance
Through–across:	force–current
	velocity–voltage
	mass–capacitance
	compliance–conductance
	mechanical resistance–conductance

The three elements all have the same velocity, so using the force–flow method they must carry the same current in the electrical analogue, therefore the electrical elements are in series. Conversely, using the through–across method, the electrical elements must have the same voltage, so they must appear in parallel.

In drawing these analogies there is a problem that a mass, unlike springs or dashpots, does not have two distinct terminals or ends, so it is not clear how one replaces it by an inductor or capacitance. However, the motion of any mass is always with respect to some reference surface, so one end of the equivalent inductor or capacitor must be grounded.

The two equivalents are shown in figure 2.6.1(a) and are redrawn in figure 2.6.1(b). The two circuits are said to be *duals*. Their elements are all interchanged but they have exactly equivalent defining equations. The two variable classifications always produce dual circuits.

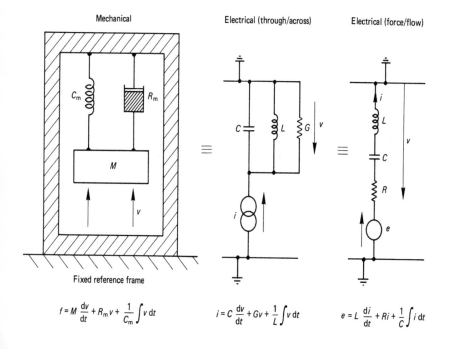

Mechanical Electrical (through/across) Electrical (force/flow)

$$f = M\frac{dv}{dt} + R_m v + \frac{1}{C_m}\int v\,dt$$

$$i = C\frac{dv}{dt} + Gv + \frac{1}{L}\int v\,dt$$

$$e = L\frac{di}{dt} + Ri + \frac{1}{C}\int i\,dt$$

Figure 2.6.1(a) *Electrical analogues of mass–spring system.*

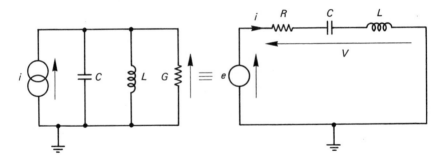

Figure 2.6.1(b)   Dual circuits of figure 2.6.1(a).

2.6.2.   The equivalent circuit is shown in figure 2.6.2(a). Most thermal
equivalents have a capacitance and resistance in parallel.
This simple analogue is not strictly accurate. Thermal capacitance and
conductance are *distributed* parameters; they cannot be represented by a
single value at a point, unlike electrical capacitance and conductance.
Figure 2.6.2(b) shows a more accurate analogy.

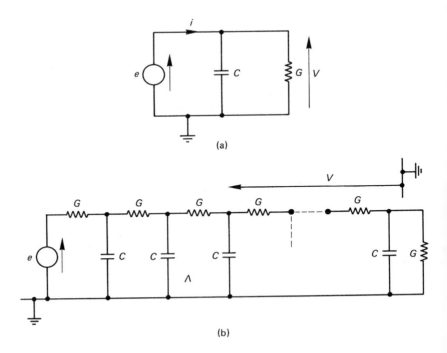

Figure 2.6.2   *'Lumped' (a) and 'distributed' (b) electrical equivalents of simple thermal circuit.*

2.6.3. (a) A lever (or a gearbox).

(b) Yes (all masses represented by capacitors with one end grounded).

(c) No (because some capacitors may have neither end grounded). It is possible to overcome this limitation, but only with considerable complication.

(d) None of them do (because there is only one storage element).

(e) It is known as 'inertance' and is proportional to liquid density, but is usually small and may often be ignored.

Solutions to chapter 3 exercises

3.5.1. (a) Types of energy (relevant to measurement): mechanical, electrical, magnetic, thermal, radiant.

Modifiers: convert between forms of same type of energy, usually between the through variable and the across variable, for example, flow transducer (flow→pressure)

Self-generators: transduce directly between two types of energy, for example, piezoelectric crystal (deformation→ charge)

Modulators: require external energy source and modulate flow of energy, for example, thermistor in bridge circuit.

(b) Self-generators:

radiant–electrical: photovoltaic effect; silicon solar cells

mechanical–electrical: electromagnetic effect; tachogenerators; piezoelectric effect; force transducers etc.

magnetic–electrical: electromagnetic effect; field measurement

thermal–electrical: thermoelectric effect; thermocouples.

Modulators: (electrical in, electrical out)

radiant: photoconductive effect; photoconductive cells

mechanical: piezoresistive and dimensional–resistive; strain gauges (semiconductor and metallic)

magnetic: magnetic–resistive; magnetic fields (little used)

thermal: thermal–resistive; resistance thermometers, thermistors.

Modifiers:

mechanical: flow/pressure; orifice-plate flow transducer

radiant: heat flux/temperature; thermal photodetector

thermal: heat flux/temperature; anemometer.

3.5.2. (a)

Photovoltaic	Radiant flow–electric current	Transforming
Pyroelectric	Temperature–charge	Gyrating
Electromagnetic	Velocity–voltage	Transforming
Piezoelectric	Displacement–charge	Gyrating
Thermoelectric	Temperature–voltage	Transforming

(b) Mechanical

Beams and springs:	Force–displacement	Gyrating
Diaphragms/tubes:	Pressure–displacement	Gyrating
Orifices etc.:	Flow–pressure	Gyrating

| Radiant Photodetectors: | Radiation–temperature | Gyrating |
| Thermal anemometers: | Heat flow–temperature | Gyrating |

Note: There is no point in doing this for a modulator since it can be arranged to produce either a current or voltage output as required.

3.5.3. Faraday's law $(e = -nd\phi/dt)$ can be used directly in measuring changing magnetic fields, as a magnetic–electric self-generator. However, it is more useful for velocity measurement as a tachometer, involving a magnet/coil system, and is then said to employ the electrodynamic effect. The effect is then that of a mechanical–electrical self-generator (at first sight it appears to be a magnetic–mechanical–electrical modulator, though in fact no magnetic energy is used). A similar effect occurs in the electromagnetic flowmeter, in which a magnetic field applied perpendicular to the flow of a conductive fluid produces an e.m.f. across electrodes in the fluid.

Faraday's law also applies to the movement of individual charge carriers in semiconductors and gives rise to the magnetoresistive effect, in which deflection of the carriers in a magnetic field leads to an increased resistance, providing a magnetic modulator (via a bridge circuit). The related Hall effect, in which an e.m.f. is produced across a strip of material in a magnetic field, can be used either for field measurement (electric–magnetic–electric modulator) or for current measurement (electric–electric modifier, since magnetic energy is not consumed).

The effect can thus be used to provide transducers of all three basic types.

Solutions to chapter 4 exercises

4.4.1. An equivalent circuit is shown in figure 4.4.1.

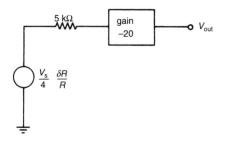

Figure 4.4.1

$V_{out} = -20 \times 4/4 \times 0.5/10$ per °C
$r = 1$ V/°C (at operating temperature).

4.4.2. Figure 4.4.2 shows the equivalent circuit. Note that a $1 : 1 + 1$

transformer is used (this is much easier to wind than the $1 : \frac{1}{2} + \frac{1}{2}$ type used in deducing the general formulae).

Charge amplifier: gain (with respect to generator) is -4

responsivity $= V_s \, \delta C / C \times 4 = 0.4$ V/mm

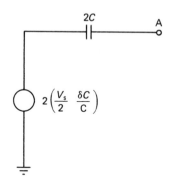

Figure 4.4.2

Non-inverting amplifier: gain 100 (strictly $10^6/(10^6 + 10^4)$)

responsivity $= V_s \, \delta C / C \times 100 = 10$ V/m

However, this requires that the amplifier has a very high input impedance, substantially greater than the impedance of the capacitors (about 100 kΩ at 100 kHz).

4.4.3. Figure 4.4.3 shows a noise-equivalent circuit of figure 4.10.

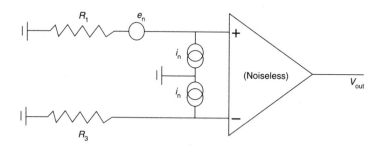

Figure 4.4.3

$$E_n = e_n (f_{ce} \ln(f_H/f_L) + f_H - f_L)^{\frac{1}{2}}$$
$$= 10^{-8}(100 \ln(10^4) + 1000)^{\frac{1}{2}} = 44 \times 10^{-8} \text{ V r.m.s.}$$

$$I_n = i_n(f_{ci}\ln(f_H/f_L) + f_H - f_L)^{\frac{1}{2}}$$
$$= 10^{-12}(1000\,\ln(10^4) + 1000) = 101 \times 10^{-12}\ \text{A r.m.s.}$$
$$V_{out} = R_2/R_1(E_n{}^2 + I_n(R_1{}^2 + R_3{}^2))$$
$$= 100(1.9 \times 10^{-13} + 1.0 \times 10^{-20} \times 10^8)^{\frac{1}{2}} = 170\ \mu\text{V r.m.s.}$$

Note: A resistor R produces white (frequency-independent) noise given by $e_n{}^2 = 4RkTB$, where k is Boltzmann's constant, T is the absolute temperature and B is the bandwidth. The two resistors appear in parallel and produce noise at the output of about 13 μV r.m.s., which is negligible. However, a 741 is a relatively noisy device, and a good low-noise amplifier would be about an order of magnitude better, so the resistor noise would have to be taken into account.

Solutions to chapter 5 exercises

5.5.1. The main advantages of LVDTs are small size, ruggedness and tolerance of environmental conditions; the main disadvantages are relatively low responsivity, phase shifts (depending on frequency) and relatively high quadrature signals. For displacements in the μm region, a device of small range (say, 1 mm) would be preferable, with responsivity, say, 10^3 or 10^4 V/m.

Variable-area capacitance transducers have fairly low responsivity of about 100 V/m, but have very low quadrature signals, permitting high gain in the following amplifier. They tend to be relatively large and can be seriously affected by moisture and other environmental factors. They are very linear, apart from edge effects near the limit of their range. They usually have a range of ±2 or 3 cm, since the capacitances would be unduly low with smaller plates.

5.5.2. (a)

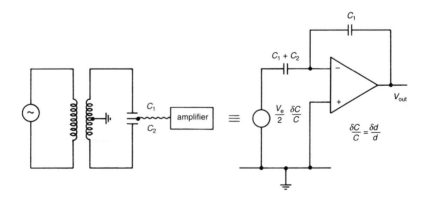

Figure 5.5.2 *Equivalent circuit of tiltmeter.*

(b) $\quad C_1 + C_2 = 2\,\dfrac{\epsilon A}{d} = 2 \times \dfrac{10^{-9}}{36\pi} \times \dfrac{10^{-3}}{5 \times 10^{-4}} = 35\ \text{pF}$

$\text{resposivity} = \dfrac{V_e}{2d} \times \dfrac{(C_1 + C_2)}{C_f} = \dfrac{2}{2 \times 5 \times 10^{-4}} \times \dfrac{35}{10} = 7000\ \text{V/m}$

A tilt of θ radians produces a displacement $L\theta$, so the responsivity in terms of tilt is 350 V/radian.

(c) The linearity will be poor unless the displacement of the bob is much less than the plate separation. A variable-area transducer would have much better linearity (though lower responsivity). Usually, however, such systems are operated in force-feedback arrangements, and the transducer is then used only as a null detector, so the linearity is much improved.

5.5.3. A resolution of 0.01 mm requires 1000 lines/cm which is rather high, and it would be easier to use gratings of 100 lines/cm and ×10 interpolation. An LED could be used as the source, with 2 or 4 photovoltaic cells attached to the 2-phase (or 4-phase) index. The separation between the gratings would have to be small (< 1 mm), so a good mechanical design is required.

Moiré gratings are the most obvious solution here, and are well suited to the application, since their main disadvantage (lack of absolute datum) is unimportant because the device would always be zeroed before each measurement.

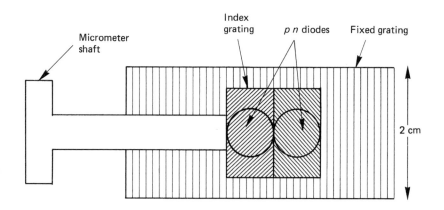

Figure 5.5.3 *Micrometer system.*

5.5.4. The terms *responsivity, detectivity, range* and *accuracy* have been discussed in the text. Accuracy (sometimes called 'precision') means the absolute accuracy (that is, the difference between 'true' value and measured value), whereas detectivity is a relative term.

(i) This can be done with a variable-separation capacitive transducer but is rather beyond the limits of the best LVDTs. The range is very small but it is impractical to use a plate separation of less than 0.1 mm. With an excitation of, say, 10 V this gives a responsivity of 10^5 V/m, so a pre-amplifier of gain 100 is required. Capacitive devices are essentially noiseless and the detectivity is set by the noise in the amplifier and can fairly easily be made equivalent to better than 10^{12}/m (per Hz).

(ii) A variable-area capacitive device or an LVDT could be used. An LVDT would be easier, being more compact and readily available for a range of 1 cm. A typical responsivity would be 1000 V/m (say, 10 V excitation). The maximum velocity of 1 cm/s requires that the measurement be performed fairly rapidly (in about 1 ms) and the excitation frequency should be about 10 times the reciprocal of this time, that is, about 10 kHz. A detectivity of 10^5/m is well within the capabilities of even the author's LVDTs.

5.5.5. (i)

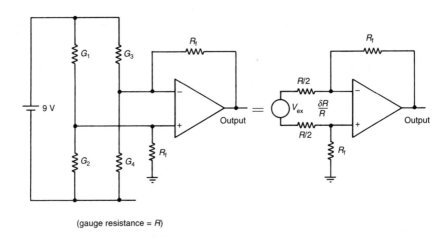

(gauge resistance = R)

Figure 5.5.5 *Equivalent circuit of strain gauge system.*

$$V_{out} = V_{ex} \times \frac{\delta R}{R} \times \text{gain}$$

$$GF = \frac{\delta R/R}{\text{strain}} \quad \text{so } V_{out} = V_{ex} \times GF \times \text{gain (per unit strain)}$$

$$= 9 \times 2 \times 200$$

$$= 3600 \text{ V/unit strain}$$

(ii) Surface stress $= \dfrac{6fx}{bd^2} = \dfrac{6f \times 4 \times 10^{-2}}{10^{-2} \times 4 \times 10^{-6}} = 6 \times 10^6 \text{ N/m}^2$ per N

$$\text{Surface strain} = \frac{\text{stress}}{E} = \frac{6 \times 10^6}{2 \times 10^{11}} = 3 \times 10^{-5} \text{ per N}$$

$$\text{Amplifier output} = 3600 \times 3 \times 10^{-5} = 0.108 \text{ V/N}$$
$$(= \text{ overall responsivity})$$

$$\text{Max. force} = \frac{9}{0.108} = 83.5 \text{ N} \left(\text{corresponding strain} \approx \frac{1}{400}\right)$$

Solutions to chapter 6 exercises

6.3.1. The temperature of the handle will be in the range 20–100°C, mostly about 30°C, and an accuracy of about 1°C would be sufficient.
  (i) Thermistors are suitable for the temperature range, though linearisation (or a look-up table) will be necessary if the range is more than about 10 °C. They have a short time constant and could easily be fitted to the handle. A simple d.c. bridge would be satisfactory, including a linearising resistor.
  (ii) Resistance thermometers could be used, the small Pt-film types being best. However, they are a little bulky for this application, though they could easily be fitted to the handle. Their resistance is low, so some compensating leads may be necessary.
  (iii) Thermocouples are usually used for higher temperatures but could certainly be used here. Most types would be satisfactory, especially Chromel/Alumel, and most commercial units would give the required accuracy. They are a little bulky and not very well suited here.
  All three transducers could be used, but thermistors are preferred.

6.3.2. 500–1000°C: Thermocouples or resistance thermometers should be used. Chromel/Alumel thermocouples feeding a commercial cold junction compensated electronics unit would be easiest.
  50–100°C: A thermocouple could be used here too, though the accuracy is near the limit of most units. Thermistors would be suitable with some linearisation or alternatively *p–n* diodes (calibration needed) or 'current sources'.
  25–35°C: A thermistor would be satisfactory here (alternatively a *p–n* diode or 'current source' device). An accuracy of ±0.1°C is well within the capabilities of thermistors, but much more difficult with thermocouples and also with *p–n* diodes (since calibration is needed). A simple d.c. bridge would suffice.

6.3.3.

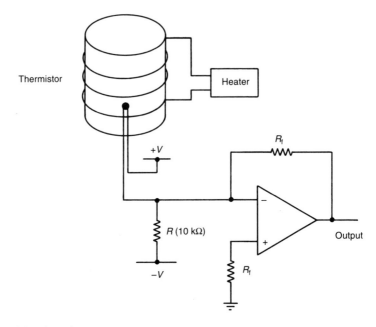

Figure 6.3.3 *Thermal system.*

If the excitation is V_s the power dissipated is $V_s^2/4R$ which must be less than 10^{-4} W so $V_s \approx 2$ V. A simple d.c. thermistor bridge will be satisfactory since a detection level of about 0.01°C is required and most good operational amplifiers (for example, 741S) will not produce excessive drift (the resistor R_1 should be equal to that seen at the inverting input, which is 10 kΩ in parallel with R_f). The output is given by

$$\frac{V_s}{4} \times \frac{\delta R}{R} \times \frac{R_f}{R/2} = \frac{1}{2} \times \frac{400}{10^4} \times \frac{R_f}{5 \times 10^3} = 1 \text{ V/°C}$$

so R_f should be 250 kΩ.

6.3.4.   Responsivity at 300 K $= \dfrac{V_s}{4} \dfrac{\delta R}{R} \dfrac{R_f}{R/2}$

$$\delta R = \frac{B}{T^2} R_T \, \delta T = \frac{3000}{(300)^2} \times 10^4 = 333 \ \Omega/\text{K} \quad \text{(thermistor alone)}$$

$$r = \frac{4}{4} \times \frac{333}{15 \text{ k}} \times \frac{10^5}{15 \text{ k}/2} = 0.296 \text{ V/K}$$

(Note that R is 15 k in this equation.)
Since $R_T = Ae^{\beta/T}$ and has a value of 10 kΩ at 300K, $A = 0.4534$.
At 310K, $R_T = 7233 \ \Omega$, so the total resistance is 12 233 Ω.

$$V_0 = \left(\frac{2}{15 \times 10^3} - \frac{2}{12\,233}\right) \times 10^5 = 2.88 \text{ V}$$

If linear, $V_0 = 0.296 \times 10 = 2.96$ V.
% error = 2.7.
At 400K, $R_T = 821 \ \Omega$ so total resistance is 5821 Ω.

$$V_0 = \left(\frac{2}{15 \times 10^3} - \frac{2}{5821}\right) \times 10^5 = 21.03 \text{ V}$$

If linear, $V_0 = 0.296 \times 100 = 29.6$ V.
% error = 18.
Note that this is a vast improvement on the linearity compared with that
of the worked example (page 87) for a simple non-linearised circuit.

Solutions to chapter 7 exercises

7.5.1. A linear device is preferable, covering the visible region and extending a
little into the infra-red, and with a fairly short time constant.
  (i) CdS. This is widely used in exposure meters because of its high
      responsivity. However, it is very non-linear, variable from device to
      device, slow, subject to drift and has a limited spectral response. In
      other words, it is not a good choice.
  (ii) Silicon solar cell. Linear (in current), good spectral response (up to
      1 μm) and very fast (much faster than required). Filters will be
      necessary to make measurements at different wavelengths, so some
      calibration will be needed (for each filter, and because the
      responsivity varies with wavelength in any case). A simple current
      amplifier is required. This is the best choice.
  (iii) Pyroelectric cell. Responds well into the infra-red, but does not
      respond at d.c. Not suitable.

7.5.2. (i) Almost any type could be used, but a cheap photoconductive type
      such as CdS is most appropriate. A very coarse switching action is
      all that is needed and the exact level is not very critical.
  (ii) A good infra-red response is essential since the peak emission will
      be at about 10 μm. Possible devices are PbS (response only to
      3 μm), InSb (spectral response to 6 μm, but may need to be cooled),

or a thermal type such as a thermistor bolometer or a pyroelectric cell (though does not respond at d.c.). The latter is probably the best choice here.

(iii) Most meters use CdS, since its responsivity is large and its spectral response is similar to that of the eye, but it is non-linear and subject to drift. A better choice would be a silicon solar cell with a suitable filter ('eye-response' silicon cells are now available and could be used).

(iv) A fast response is required here and a silicon solar cell is the obvious choice, being linear in current responsivity and suitable in spectral response. A simple current amplifier can be used.

(v) A high detectivity is needed and a device suited to low light levels would be best, such as a photomultiplier or a PIN photodiode (or avalanche photodiode). PbS would be suitable if a response into the infra-red was needed, but a PIN diode would probably be the best choice.

7.5.3. Diameter of beam at detector $\approx 2 \times \dfrac{20}{60}$ so area $\approx 3 \times \dfrac{4}{9} \times \dfrac{1}{4} \approx \dfrac{1}{3} \, m^2$

$$\text{Radiation received} \approx 10 \times \frac{10^{-4}}{0.33} \approx 3 \times 10^{-3} \, \mu W$$

Signal current $= rW = 3 \times 10^{-10} A$
Amplifier output $= 3 \times 10^{-10} \times 10^5 = 3 \times 10^{-5} \, V$
This is a small signal but is fairly easily detected in a bandwidth of, say, 100 Hz, and a signal-to-noise ratio of more than 10 would be expected in practice.

7.5.4. A simple circuit is shown in figure 7.5.4(a).
For radiation of 1 kW/m^2, that received by the cell will be $10^3/10^4$ so the cell current at wavelength 1 μm will be $10^3/10^4 \times 0.1 = 10^{-2} A$. A value of R_f of 1 kΩ will produce the required responsivity of 10 V per kW/m^2. Figure 7.5.4(b) shows the black body curve and the cell response.
In order to obtain reasonable proportionality between the cell output and the radiant energy, the two curves should have similar shape, and a filter that starts to attenuate above 0.5 μm can be used, shown by the broken line in figure 7.5.4(b). This will affect the design of the circuit in figure 7.5.4(a), and R_f will have to be increased. The exact value is not easy to calculate and calibration (against a thermal detector) will be necessary. Such a detector could be used for this application, of course, but a solar cell with a filter is much more convenient.

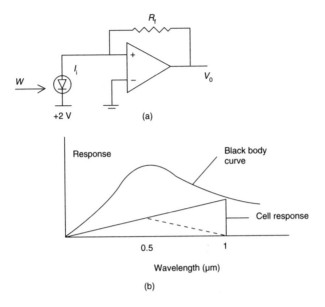

Figure 7.5.4 *(a) Simple circuit for solar radiation monitor. (b) Black body curve and cell response.*

Solutions to chapter 8 exercises

8.6.1. The transfer functions are

$$\text{(a)} \quad \frac{x_r}{\ddot{x}_{in}}(s) = \frac{1}{s^2 + 2\xi\omega_0 s + \omega_0^2}$$

$$\text{(b)} \quad \frac{v_r}{v_{in}}(s) = \frac{s^2}{s^2 + 2\xi\omega_0 s + \omega_0^2}$$

(i) Absolute velocity (10 Hz to 1 kHz)
Using (b) the response is flat (and of value unity) above the natural frequency f_0. We need to measure relative velocity v_r in a system with natural frequency less than 10 Hz (to avoid the peak produced at f_0). A simple cantilever beam can be used with a magnet/coil sensing system (better with a fixed magnet).
(ii) Absolute acceleration (10 Hz to 10 kHz)
Using (a) the response is flat (and of value $1/(\omega_0^2)$) below the natural frequency. We need to measure relative displacement x_r below f_0, so f_0 should be greater than 10 kHz. A piezo-accelerometer is necessary to achieve this high value of f_0, and does not operate down to d.c. when used as a displacement sensor. The design of the charge amplifier will determine the lower cut-off frequency.

(iii) Absolute displacement (0–100 Hz)

Equation (b) can be written in terms of x_r and x_{in} and permits absolute displacement to be measured (using a relative displacement transducer) above f_0. The only way to obtain absolute acceleration from d.c. upwards is to use relation (a) and double-integrate the acceleration. An open-loop system with $f_0 > 100$ Hz could be used, but drift problems are severe (because of the double integration) and closed-loop devices are preferable. A force-feedback accelerometer with closed-loop resonant frequency > 100 Hz (open-loop $f_0 \approx 10$ Hz) would be used.

8.6.2. (i) Compliance $=$ deflection/force $= \dfrac{4l^3}{Ebd^3} = 2.5 \times 10^{-4}$ m/N

$$\omega_0^2 = \frac{1}{MC} = 4 \times 10^4, \qquad f_0 = 31.8 \text{ Hz}$$

(ii) The equivalent circuit is shown in figure 8.6.2.

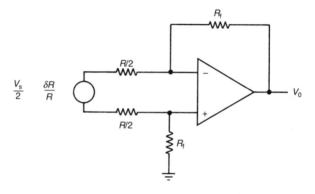

Figure 8.6.2

$$V_0 = \frac{V_s}{2} \times \frac{\delta R}{R} \times \frac{R_f}{R/2} = \frac{V_s}{2} \times 2 \times \frac{R_f}{R/2} \times \text{strain} = 2000 \times \text{strain}$$

Below resonance $\dfrac{x_r}{\ddot{x}_{in}} = \dfrac{1}{\omega_0^2}$, so $x_r = \dfrac{1}{\omega_0^2}$ per m s^{-2}

$$\text{Strain} = \frac{\text{stress}}{E} = \frac{6fx}{Ebd^2} = 12 \times 10^{-5} f \qquad \left(f = \frac{x_r}{C} \right)$$

$$V_{out} = \frac{2000 \times (0.48)}{4 \times 10^{-4}} = 0.024 \text{ V/m s}^{-2} \ (= \text{ responsivity})$$

(iii) ± 10 V $= 0.024 \times a_{max}$ so $a_{max} = 400$ m s$^{-2} \approx 40\,g$
(the deflection produced ≈ 1 cm, which is rather high).

8.6.3. The compliance = C_m

$$\frac{4l^3}{Ebd^3} = \frac{4 \times 125 \times 10^{-6}}{2 \times 10^{11} \times 2 \times 10^{-2} \times 64 \times 10^{-9}} \approx 2 \times 10^{-6} \text{ m/N}$$

Deflection for force of 1000 N $\approx 2 \times 10^{-6} \times 10^3 \approx 2$ mm

(i) Strain gauges. Max. surface stress =

$$\frac{6fx}{bd^2} = \frac{6 \times 10^3 \times 2 \times 10^{-3}}{2 \times 10^{-2} \times 16 \times 10^{-6}} \approx 0.4 \times 10^8 \text{ N/m}^2$$

$$\text{Max. strain} = \frac{0.4 \times 10^8}{2 \times 10^{11}} \approx 2 \times 10^{-4}$$

Strain gauges are very suitable here. They have no particular disadvantages, apart from temperature effects which can be reduced by using several gauges as discussed in the text. Metallic foil types would be the best solution.

(ii) Piezo-crystals are not suitable here, as they do not respond to displacement at d.c., and a d.c. system is specifically stated.

(iii) Variable-separation capacitive transducers could certainly be used, though a rather wide plate spacing is needed (> 2 mm). However, it would be much more complicated to attach them than strain gauges, and they are a very poor choice here.

8.6.4. (i) This can be done by a laser Doppler system or by a correlation method – in either case it is going to be rather expensive. The laser system would be appropriate if there was a very wide flow range (up to high values) and if high accuracy is needed. Otherwise a correlation system, involving two separated beams of light passed through the tube, would be appropriate. The photodetector outputs can be digitised into binary values for easy interfacing and fast computation.

(ii) The rate of flow will vary with depth but it should be assumed that an average value is required. A 'weir' is probably the simplest method, measuring the height difference produced (rate $\approx$ (height)$^{1.5}$). If such an obstruction is not permissible a turbine could be used, adjusting the depth to give an average value. Alternatively, if no contact or disturbance is possible an optical correlation method could be used, though it would measure the velocity at the surface.

(iii) Small turbines in short tubes could probably be best here, since they could easily be placed at any depth and could also provide directional information by orienting the cylinder for maximum flow. A simple proximity detector providing a digital readout could be used. Other possible transducers are orifice plates (in cylinder), though some disturbance in flow will occur, or drag-force devices.

Solutions to chapter 9 exercises

9.5.1. The power in the coil is given by $I^2 R$. Thus $I^2 \times 7 = 50$ and $I = 2.67$ A. The force is given by $(Bl)I$ and the maximum force is therefore $10 \times 2.67 = 26.7$ N.

9.5.2. The force due to both springs at full-scale deflection is $2.5 \times 10^{-7} \times 2 \times 2 = 10^{-6}$ N m. The current required to produce this force is 10^{-6} N m/0.01 N m/A $= 10^{-4}$ A.

9.5.3. A speed of 15 000 rpm $= (15\,000 \times 2 \times \pi)/60$ radians/s. The motor constant k_m is given by voltage divided by angular speed, thus $k_m = 6/1571 = 3.82 \times 10^{-3}$ V/radian/s. The torque of the motor is given by $k_m \times I$ which gives a maximum torque of $3.82 \times 10^{-3} \times 3 = 11.46 \times 10^{-3}$ N m.

9.5.4. The force of attraction between the plates is given by

$$F = \frac{\epsilon_0 A V^2}{2d^2}$$

The force F is thus $8.85 \times 10^{-12} \times 50 \times 10^{-6} \times 50^2/\{2 \times (10 \times 10^{-6})^2\}$ N which gives a force of 5.53×10^{-3} N.

Solutions to chapter 10 exercises

10.6.1. (a) The main advantages of optical-fibre sensors are their immunity from electromagnetic interference and the ease with which they can be linked into an optical communication system. There are no particular disadvantages, though many of the sensing applications are somewhat artificial, the fibre being used essentially as a light guide. One interesting development is that the power for a sensing system can sometimes be fed down the fibres themselves.

The main properties of light suitable for such sensors are intensity, phase, polarisation, wavelength and spectral distribution, and the main modulating effects are piezo-absorption, stress bi-refringence, an electro-optical effect, a magneto-optical effect, the Faraday effect and some thermal effects.

The most widely used devices are intensity modulators, often employing the relative motion of two fibres, and sometimes using a 'microbending' arrangement (piezo-absorption) to measure force or other parameters derived from it. Phase modulators are not much used as sensors, though the fibre-optic gyroscope uses this effect. However, polarisation modulators, using Faraday rotation, are a valuable non-contacting means of measuring large currents.

(b) Optical processing involves transforming and manipulating optical data by optical means, a particularly good example being afforded by an optical Fourier Transform whereby the light distributions in the front and back focal planes of a lens are a Fourier pair. Such systems were developed well before the recent microelectronics developments but suffered from difficulties in input and output of data. Indeed, the most successful applications (deblurring photographs and character recognition) did not use real-time data. To some extent the development of microprocessors has reduced the investigation of optical processing, since it is now relatively easy to digitise a picture and carry out a Fast Fourier Transform. However, fundamental limits in size and speed of silicon-based devices are gradually being reached and optical systems, which have inherently very high data rates, may then receive a new impetus. 'Optical computers' are certainly a long way off, since many of the basic components simply do not exist, but there is sure to be a valuable 'spin-off' from developments in communication systems. The 'direct' coupling of fibres to conventional computers is relatively easy, of course, and more and more optical processing will be carried out before the data are transferred to the computer, which is a first step towards a complete optical processor.

10.6.2. (a) Advantages: freedom from electrical interference (serious problem with capacitive devices); remote measurement; no interaction with object measured; no leads near object measured; small size

  Disadvantages: drift in reading due to changes in light source; small size may be disadvantage in some cases where averaging over area is needed; steady light level leads to inherent shot noise limit, capacitive devices are inherently noiseless

  (b) Responsivity: slope $= 10^{-4}$ W/0.8 mm $= 0.125$ W/m
  output from photodiode $= 1.25 \times 10^{-2}$ A/m

  Detectivity : photocurrent at C $= 5 \times 10^{-6}$ A

$$\text{Shot noise}: i_m^2 = 2eI\Delta f$$

$$= 2 \times 1.6 \times 10^{-19} \times 5 \times 10^{-6} \times 10^2$$

$$= 1.6 \times 10^{-22} \text{ A}^2$$

$$i_{rms} = 1.3 \times 10^{-11} \text{ A}$$

  Least detectable displacement: $1.25 \times 10^{-2} \times d = 1.3 \times 10^{-11}$
  $d = 10^{-9}$ m   detectivity $= 10^9$/m

10.6.3. The particular advantages of resonator sensor systems are that they produce a frequency output proportional to the quantity of interest, suitable for interfacing to computer systems, and that their stability is determined by mechanical components and is therefore high and little affected by electrical interference.
  Several physical parameters can be measured by such systems. Devices may be classified into wires, beams and cylinders. Wire systems were the

first to be used, usually for measurement of force or pressure. Vibrating beams may similarly be used for force measurement, but can conveniently be used for liquid density, level or viscosity by placing a part of the beam in the liquid of interest. It is also possible to measure flow by using such a beam in a tube. Very small beams made from quartz or etched from silicon can be produced, often in a double tuning-fork arrangement, and are used in miniature pressure or vibration sensors. Vibrating cylinders are similar to vibrating beams, but are well suited to density and flow measurement by designing a suitably vibrating section of tube.

A rather different principle is used for the detection of some chemical quantities, a vibrating quartz crystal having a suitable absorbing film deposited on one face, and the change in frequency monitored.

It is often suggested that a resonator sensor system is inherently superior to a conventional analogue system, because it is 'inherently digital'. In fact it is nothing of the sort, but is simply of a form that is easy to digitise accurately. However, there are severe costs to be paid for this, in the form of inherent non-linearity and severe temperature dependence, and in many cases one would be better off with a conventional analogue system and a simple ADC unit. The most promising devices are the miniature beams and quartz crystals with absorbing layers.

10.6.4. (a) The properties of silicon that make it suitable for VLSI are that it can be produced in very pure large crystals and that its oxide provides an excellent mask for selective doping and etching. These properties make it excellent both for the production of large numbers of very small circuit elements and for the physical construction of small structures for use as resonator (or other) sensors. In addition, there are a number of useful effects which can be used for self-generators or modulators.

The main self-generator effects are the Seebeck, photovoltaic and galvano-electric. It lacks a suitable mechanical or magnetic effect, but because of the ease of masking and doping, a layer of suitable material can easily be added. There are modulating effects for all the main energy forms, the most important being piezoresistive, resistance/temperature, photoconductive and magnetoresistive/Hall. The use of silicon-based solid-state sensors will certainly increase substantially in the next few years.

(b) The premises on which the theory is based appear very reasonable. Carbon forms an enormous number of compounds and appears to be the only element capable of forming the complicated molecules necessary for the natural development of organic life. Although silicon-based compounds could not develop naturally in this way, silicon is a remarkably good material for producing miniature electrical circuits, as discussed above. It seems likely that as packing densities are increased, units will be produced that considerably surpass the storage of the human brain, and with an access time many orders of magnitude shorter. Of course, what constitutes intelligence and whether it can ever be truly artificial is another matter, and not within the intended scope of this book. However, to be on the safe side, the writers' new year resolutions are to be kinder to their home computers!

10.6.5.

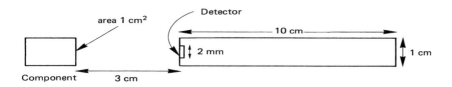

Figure 10.6.5(a) *Radiant system.*

Radiation received $W = \dfrac{\epsilon\theta a_1 a_2 T^4}{\pi d^2}$ where $\epsilon =$ emissivity (say, 0.5)

$$(\text{assume } T \approx 300\text{K}) = \frac{1}{2} \times \frac{6 \times 10^{-8} \times 10^{-4} \times \pi \times 4 \times 10^{-6}}{\pi \times 9 \times 10^{-4} \times 4} \times 81 \times 10^8$$

$$\approx 27 \ \mu\text{W}$$

Change per °C $= \delta W = 4W/T \times \delta T = 36 \times 10^{-8}$ W/K

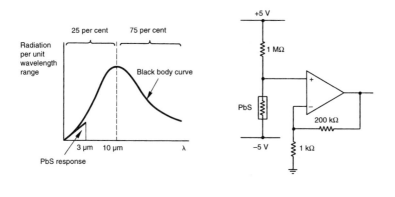

(b)   (c)

Figure 10.6.5(b), (c) *Black body curve (b), and detector/amplifier (c).*

The detector could be PbS (3 μm max.), InSb or pyro-electric (a.c. only). PbS would be most convenient, in a simple d.c. arrangement. The peak radiation will be at 2900/T μm ≈ 10 μm, so PbS will intercept only about 1/12 of the total radiation and its mean effective responsivity will be about 5 per cent of its peak value say, about 500 V/W.

The output from the amplifier will be $36 \times 10^{-8} \times 500 \times 200 = 36 \times 10^{-3}$ V/°C. In practice a responsivity of, say, 0.1 V/°C would be appropriate, so a little more gain is needed. The system should be able to detect a change of about 1°C and this can be achieved fairly easily with PbS.

10.6.6. (a) (i) Received power: $(\alpha_{qs} \times \alpha_{sw} e^{-al})^2 \times \alpha_{wa}$
$R = pc$: $R_q = 13.7$, $R_s = 48.8$, $R_w = 1.5$, $R_a = 0.44 \times 10^{-3}$ ($\times 10^6$)
$\alpha = 4R_1R_2/(R_1 + R_2)^2$: $\alpha_{qs} = 0.68$, $\alpha_{sw} = 0.112$, $\alpha_{wa} = 1$
Received power = $(0.68 \times 0.112 \times 0.61)^2 \times 1 = 2.16 \times 10^{-3}$
Round trip (air only): $t_r = (2 \times 0.1)/(1.5 \times 10^3)$
$= 1.33 \times 10^{-4}$ s
(ii) Make pulse width = $0.1 \times t_r \approx 10$ μs
repetition time = $10 \times t_r \approx 1$ ms
(Assume that attenuations in quartz and steel are zero, ignore time to pass through pipe wall, assume 100 per cent reflection at water–air interface).
Resolution ≈ crystal period = 1 μs
1 μs: 133 μs gives about 1 mm
(b) (i) Frequency shift = $2f/C \cos \theta v$
$= 2 \times 10^6/(1.5 \times 10^3) \times 0.87 \times 1$
$= 1.16$ kHz
This is an easily measurable change.
(ii) Transit time methods suffer from low signal levels (because of the extra boundary) and very small time changes. The water velocity is less than 1/1000 of the sound velocity so the change in time will be of this order, that is, 0.1 μs.

Solutions to chapter 11 exercises

11.5.1. (a) *Resolution* is the smallest measurable change in the input in a given time, for example 1 part in 10^6 in 1 second.
Accuracy is the difference between the measured value and the true value.
Range is the total range of values over which the system operates.
Measurement time is the averaging time over which a measurement is made.
(b) A frequency counter operating with an internal clock of 1 MHz and operating in period mode should be used. In order to obtain a measurement time of 0.1 s the input frequency must be doubled and the gate opened for the time taken for ten periods. The resolution

will be 1 in 10^5 in 0.1 s and the recorder should monitor only the second and third least significant digits to give 0.5 Hz full scale.

11.5.2. (a) There are three basic classes: frequency output devices, time output devices and inherently digital devices.

Frequency output: These may be analogue transducers operated so as to produce a frequency output (such as a capacitive device in an oscillator circuit) or a resonator sensor.

Time output: These are similarly analogue devices operated to produce a time interval, similar to some types of ADC.

Inherently digital: These are devices such as encoders, gratings or interferometric systems, in which the fundamental operation is digital.

The required resolution or dynamic range can be designed into the system and usually leads to better results with better temperature compensation etc.

(b) (i) This could be done with gratings or by interferometry, but a resistive or capacitive method producing a frequency or time interval would be cheaper. A grating system with interpolation to obtain the required resolution would be best.

(ii) A thermistor system is probably needed to obtain the required resolution, which is at the limit for thermocouples; it could be done with a *p–n* junction. A thermistor in an *RC* oscillator can be arranged to cancel most of the non-linearity, and is the best solution, though a time-interval method could also be used.

(iii) A thermal detector could be used in an oscillator circuit to produce a frequency or time interval, alternatively a photo-conductive device. However, a *p–n* photodiode would be best, and the current produced could feed directly into an integrating circuit to produce a time interval.

11.5.3. (a) As stressed at the beginning of this book, digital transducers employ analogue effects, so the advantages of digital transducers are more concerned with the transmission and processing of the data than with its capture. The main advantages are therefore in noise-free transmission and in convenience for computation.

(b) (i) With, say, a 5 cm radius a rotation of 10 seconds produces a deflection of $1/60 \times 1/60 \times 1/6 \times 5 \times 10^{-2}$ m, requiring 432 000 lines/m which is very high so that interpolation by a factor of 10 is needed.

(ii) A magnet–coil system with a counter is needed. Timing over one revolution in period mode with a 1 MHz clock produces a count of 10^5 so the resolution is 1 in 10^5 in 0.1 s. The signal-to-noise ratio must be fairly high to avoid triggering errors.

11.5.4. Decoder: Enabled when the address is in the range $C000 to $C0FF and activates an output line determined by (S2,S1,S0). The upper four lines are write and the lower four read.

DAC: Enabled by writing to $C003.

Digital outputs: Enabled by writing to $C002, must be latched.

Digital inputs: Read from $C001, tri-state buffer needed.

ADC: Started by writing to $C001. The clock is divided down to a suitable rate for the ADC and the EOC signal disables the divider. The output enable is activated by reading from $C000 but a suitable time must be allowed for conversion.

(i)   LDA $FE, STA $C003

(ii) STA $C001, NOP as needed, LDA $C000

(iii) LDA $11, STA $C002

(iv) LDA $C001, CMP $7F, BRANCH

11.5.5. The IEEE 488 is a standardised parallel interface for connecting instruments to a controlling computer. There are 24 lines: 8 data, 5 management, 3 data transfer and 8 ground. Instruments may be talkers or listeners or both. A 3-line handshake system permits several instruments to be on the bus together. Instruments can indicate a need for service (SRQ) and can be polled in serial or parallel. The main advantages are that it is a standardised system, so devices can easily be connected. The main disadvantages are that it adds to the cost of an instrument and is only an 8-bit system.

Network measurement

Synthesiser: Need only be a talker but remote/local operation required and must be programmable in frequency and amplitude.

DVM: Must be listener and talker and programmable in range.

Printer: listener only

Flow diagram

(i)   Step through amplitude response at 1 Hz, check linearity.

(ii) Select amplitude in linear region, step through frequency response from 0.1 Hz in ratios of say 1.1, determine approximate peak.

(iii) Step linearly through frequency in 0.1 Hz intervals over range of, say, ±0.5 Hz near peak, determine actual peak.

(iv) Step through amplitude response at frequency response peak.

(v)   Print out values.

11.5.6. A smart instrument is one whose operation is controlled by an on-line microprocessor, so that errors due to non-linearity, temperature etc. can be automatically removed. The instrument is under total microprocessor control and any malfunction can be immediately notified.

Displacement transducer: This has a very small range (0.1 mm) and needs a detectivity of 1 μm. It can be achieved with small LVDTs or capacitive devices, but not with resistive transducers.

Rotary transducer: Could be absolute or incremental. An absolute encoder would be better, but an incremental device with a track producing a signal at the start of each revolution would be satisfactory. The system could be calibrated over one revolution by using a sheet of known thickness. The data could be held in a look-up table and used to modify the LVDT data in accordance with the position of the roller, whose bearings and shape will produce repeatable errors on rotation. The system is feasible, and such systems have been used.

References and Bibliography

It would not be very difficult to produce an enormous list of references, because almost every physical effect or transducer mentioned could be separately referenced. However, it was thought to be more useful to give specific references to only some of the more recent developments, but to provide a general bibliography to the subject as a whole. Most of the books mentioned are of course books on measurement, since there are very few available that concentrate on sensors and transducers.

References

Beck, M.S. (1983). 'Cross correlation flowmeters', *Instrument Science and Technology,* Vol. 2 (ed. B. Jones), Adam Hilger, Bristol, pp. 89–106.

Culshaw, B. and Dakin, J. (Eds.) (1989). *Optical Fiber Sensors: Systems and Applications,* Artech House, Norwood, Massachusetts.

Doebelin, E.O. (1966). *Measurement Systems: Application and Design,* McGraw-Hill, Maidenhead, Berks.

Fish, P.J. (1993). *Electronic Noise and Low Noise Design,* Macmillan, London.

Jones, B.E. (1977). *Instrumentation, Measurement and Feedback,* McGraw-Hill, Maidenhead, Berks.

Jones and Philp (1983). *Proc. First Conf. Sensors and their Applications,* Institute of Physics, Bristol.

Middelhoek, S. and Noorlag, D.J.W. (1981a). 'Silicon microtransducers', *J. Phys. E,* Vol. 14, 1343–52.

Middelhoek, S. and Noorlag, D.J.W. (1981b). 'Three-dimensional representation of input and output transducers', *Sensors & Actuators,* Vol. 2, 29–41.

Shearer, J.L., Murphy, A.T. and Richardson, H.H. (1971). *Introduction to System Dynamics,* Addison-Wesley, London.

Spooncer, R.C. *et al.* (1987). 'Fibre optic sensors II', *Proc. SPIE*, Vol. 798, 137–41.

van Dijck, J.G.R. (1964). *The Physical Basis of Electronics*, Centrex, Eindhoven/ Macmillan, London, pp. 31–44.

Bibliography

Barney, G.C., *Intelligent Instrumentation*, Prentice-Hall, Englewood Cliffs, New Jersey, 1985.
Many examples of practical applications of computer-controlled devices.

Bentley, J.P., *Principles of Measurement Systems*, Longman, London, 1995.
Includes fundamentals and applications of many types of measurement system.

Concise Encyclopaedia of Measurement and Instrumentation (eds. L. Finkelstein and K.T.V. Grattan), Pergamon, Oxford, 1994.

Culshaw, B., *Optical Fibre Sensing and Signal Processing*, Peter Peregrinus, Hichin, Herts., 1986.
Basic principles and applications of optical fibre systems.

Doebelin, E.O., *Measurement Systems: Application and Design*, McGraw-Hill, Maidenhead, Berks., 1990.
Very comprehensive coverage of measurement systems, excellent reference book.

Gardner, J.W., *Microsensors*, Wiley, Chichester, 1994.
Principles, applications and recent developments of microsensors.

Handbook of Measurement Science, Vols 1 and 2 (ed. P.H. Sydenham), Wiley, Chichester, 1982.
Thorough review of theoretical fundamentals (Vol. 1) and applications (Vol. 2).

Hauptmann, P., *Sensors,* Prentice-Hall, Englewood Cliffs, New Jersey, 1991.
Many types of sensor, including silicon, optical fibre and chemical.

Jones, B.E., *Instrumentation, Measurement and Feedback*, McGraw-Hill, Maidenhead, Berks., 1977.
Integrated account of the measurement field, including feedback instruments.

Morris, A.S., *Principles of Measurement and Instrumentation*, Prentice-Hall, Englewood Cliffs, New Jersey, 1988.

Neubert, H.K.P., *Instrument Transducers*, Oxford University Press, 1975.
Detailed design criteria of various tranducers, especially inductive ones.

Optical Fiber Sensors: Systems and Applications (eds. B. Culshaw and J. Dakin),
Artech House, Norwood, Massachusetts, 1989.
Collected papers on OFS, much valuable material.

Sensors Series, Institute of Physics, Bristol.
Collected papers presented at the bi-annual conferences on 'Sensors and their
Applications', from 1983 to 1993.

Sydenham, P.H., *Transducers in Measurement and Control*, Adam Hilger,
Bristol, 1985.

Woolvet, G.A., *Transducers in Digital Systems*, Peter Peregrinus, Hitchin, Herts.,
1977.
Useful survey of digital devices, including encoders and resonator sensors.

Index